Pitman Research Notes in Mathematics Series

Submission of proposals for consideration

Suggestions for publication, in the form of outlines and representative samples, are invited by the Editorial Board for assessment. Intending authors should approach one of the main editors or another member of the Editorial Board, citing the relevant AMS subject classifications. Alternatively, outlines may be sent directly to the publisher's offices. Refereeing is by members of the board and other mathematical authorities in the topic concerned, throughout the world.

Preparation of accepted manuscripts

On acceptance of a proposal, the publisher will supply full instructions for the preparation of manuscripts in a form suitable for direct photo-lithographic reproduction. Specially printed grid sheets are provided and a contribution is offered by the publisher towards the cost of typing. Word processor output, subject to the publisher's approval, is also acceptable.

Illustrations should be prepared by the authors, ready for direct reproduction without further improvement. The use of hand-drawn symbols should be avoided wherever possible, in order to maintain maximum clarity of the text.

The publisher will be pleased to give any guidance necessary during the preparation of a typescript, and will be happy to answer any queries.

Important note

In order to avoid later retyping, intending authors are strongly urged not to begin final preparation of a typescript before receiving the publisher's guidelines and special paper. In this way it is hoped to preserve the uniform appearance of the series.

Longman Scientific & Technical
Longman House
Burnt Mill
Harlow, Essex, UK
(tel (0279) 26721)

Immanuel M. Bomze

University of Vienna.

A functional analytic approach to statistical experiments

Longman
Scientific &
Technical

Copublished in the United States with
John Wiley & Sons, Inc., New York

Longman Scientific & Technical,
Longman Group UK Limited,
Longman House, Burnt Mill, Harlow
Essex CM20 2JE, England
and Associated Companies throughout the world.

Copublished in the United States with
John Wiley & Sons, Inc., 605 Third Avenue, New York, NY 10158

© Longman Group UK Limited 1990

First published 1990

AMS Subject Classification: (main) 62BXX, 62CXX, 62CXX
                               (subsidiary) 62A05, 62G05, 46B30

ISSN 0269-3674

British Library Cataloguing in Publication Data
Bomze, Immanuel M.
  A functional analytic approach to statistical experiments.
  1. Statistical mathematics
  I. Title II. Series
  519.5

ISBN 0-582-06869-X

Library of Congress Cataloging-in-Publication Data
Bomze, Immanuel M., 1958–
  A functional analytic approach to statistical experiments/I.M. Bomze.
  p. cm. — (Pitman research notes in mathematics series, ISSN 0269-3674 ; 237)
  Includes bibliographical references (p.   ).
  1. Experimental design. 2. Mathematical statistical. I. Title. II. Series.
QA279.B65 1990   90-6105
519.5—dc20   CIP

Printed and bound in Great Britain
by Biddles Ltd, Guildford and King's Lynn

To my family

Preface

This work would not exist without two admired teachers of mine: it was initiated by *H. Strasser*'s valuable suggestions and received stimulation by *L. Schmetterer*'s non-ceasing, encouraging interest. My wife, *S. Bomze-de Barba*, my father *J. Bomze*, as well as my friends and colleagues, *K.A. Fröschl, W. Grossmann, H. Luschgy, B.M. Pötscher, W. Schachermayer, E. Siebert*, and last but not least *M.R. Uchida*, created through their useful comments, efficient support, and fruitful discussions, a fertile atmosphere favouring the emergence of this book. I am deeply indebted to all of them.

Vienna, January 1990 *I.M. Bomze*

Contents

5. Invariance

Introduction

Motivated by a permanently and rapidly increasing range of applications, statistical reasoning has developed a huge variety of methods. To sharpen insight and to enable easy innovation of new techniques, structurally and hierarchically organized theories seem to be indispensable. The main purpose of this book is to contribute to this goal by investigating some structural properties of statistical experiments irrespective of the complexity of the underlying model.

Statistical inference mainly deals with the following situation: the researcher is faced with a model consisting of (i) a measurable space $(\Omega, \mathcal{F})$ where Ω is the sample space and $\mathcal{F}$ represents the system of random events in Ω, and (ii) a set $\mathcal{P}$ of probability measures on $(\Omega, \mathcal{F})$. Now she or he has to decide which of the probability distributions P or, equivalently, which of the probability spaces $(\Omega, \mathcal{F}, P)$, $P \in \mathcal{P}$, is well supported by the data observed. This is the reason why the triplet $\mathcal{E} := (\Omega, \mathcal{F}, \mathcal{P})$ is called a "(statistical) experiment".

In some situations the set $\mathcal{P}$ has a property which allows us to study only one probability space $(\Omega, \mathcal{F}, Q)$. Consider the system of $\mathcal{P}$-null sets

$$\mathcal{N} := \{N \in \mathcal{F} : P(N) = 0 \text{ for all } P \in \mathcal{P}\}.$$

$\mathcal{E}$ (or, equivalently, $\mathcal{P}$) is called "dominated" if there exists a probability measure Q defined on $(\Omega, \mathcal{F})$ such that an event $N \in \mathcal{F}$ belongs to $\mathcal{N}$ if and only if its Q-probability vanishes, i.e. $Q(N) = 0$. This type of experiments is relatively easy to cope with by measure theoretic means, since most methods of measure theory dealing with a single measure space are not too complicated. Consequently, most of the results presented here have counterparts which are well known and familiar in theoretical statistics, but apply only to dominated experiments.

However, in many cases relevant to applications, $\mathcal{P}$ has, roughly speaking, too many different kinds of elements to be dominated (for an exact criterion see, e.g. [Berger 1951]). This means that the structure of $\mathcal{N}$ is too complicated to admit a simple measure theoretic treatment of most interesting problems. To overcome these difficulties, several generalizations of the domination property were introduced and investigated, for example Σ-finite [*Fell* 1956] (see also [*Fremlin* 1978] and [*LeCam* 1986, pp.13,667]), compact [*Pitcher* 1965], discrete [*Basu/Ghosh* 1967], weakly dominated [*Mussmann* 1972], coherent [*Hasegawa/Perlman* 1974], semi-decomposable [*Luschgy/Mussmann* 1986], and majorized [*Ramamoorthy/Yamada* 1983] experiments (cf. [*Diepenbrock* 1971]; this list shall only demonstrate the diversity of this approach and most likely is incomplete). Although these concepts yielded many fruitful results, they allow for a theory applicable only to restricted though vast classes of experiments. Hence – initiated by *LeCam*'s important article [1964] – the idea spread to employ functional analytic tools in the investigation of general statistical experiments, with the aim to simplify or even to replace measure theoretic arguments. One branch of this approach combines these with the investigation of functional analytic properties of stochastic kernels. The fecundity and practicability of this method culminating in a theory of statistical experiments with an emphasis on decision theory, is excellently illustrated by *Heyer*'s [1982] monograph. A different direction of research, which will be treated in this book, focuses on order theoretic arguments in the context of *Banach* lattices, by enlarging the class of objects usually investigated by statisticians. With this generalization, many statements, which without any further asssumptions on $\mathcal{P}$ are false in the classical setting, remain valid in a wider sense.

In a nutshell, *LeCam*'s methods are based on the following arguments: to enforce functional analytic methods, one has to embed the set $\mathcal{P}$ into a linear (and topological) structure, the so-called "L-space" $\mathcal{L}$ of the experiment. The topological dual space $\mathcal{M} = \mathcal{L}^*$ of $\mathcal{L}$ is named its "M-space". Thus, every $f \in \mathcal{M}$ is a (continuous) linear map from $\mathcal{L}$ to $\mathbb{R}$, which due to $\mathcal{P} \subset \mathcal{L}$ a fortiori assigns a real number to every probability measure $P \in \mathcal{P}$, and which is interpreted as a generalized random variable.

That this generalization is not too abstract, is supported by the following observation: the procedure above consists, roughly speaking, in replacing the maps $\varphi : \omega \mapsto \varphi(\omega)$ defined on sample points $\omega \in \Omega$ by maps $\tilde{f} : P \mapsto \tilde{f}(P)$ defined on probability measures $P \in \mathcal{P}$. This transition is motivated by the argument that in "reality" one does not perform point measurements $\varphi(\omega)$ but rather one averages instead, i.e. one considers integrals

$$\int_{\Omega} \varphi(\omega)\, P(d\omega) \equiv \int \varphi\, dP.$$

By this approach, one may be encouraged to study maps of the form

$$\dot{\varphi} : P \mapsto \int \varphi\, dP$$

$$\mathcal{P} \to \mathbb{R}$$

which in turn are special cases of the maps $\tilde{f} : \mathcal{P} \to \mathbb{R}$ mentioned above. Now one can show that every map $f \in \mathcal{M}$ can be obtained by a limit procedure from maps $\dot{\varphi}$. In other words, the space $\mathcal{M}$ of generalized random variables f is the completion of the space of maps $\dot{\varphi}$ obtained by averaging "classical" (bounded) random variables φ.

The power of the method sketched above is illustrated by the wide range of problems covered by *LeCam*'s monograph [1986]. See also [*Strasser* 1985] for an excellent survey which reveals the importance of these and related methods, especially for decision-theoretic questions as well as for asymptotic theory. However, in this book – which is based upon the author's thesis [*Bomze* 1987] – a somewhat different point of view is taken towards the basic notions and results occurring in both books cited above and of course in the original literature. As a consequence, new proofs and characterizations are provided which may faciliate the approach to a functional analytic treatment of statistical experiments. Furthermore, we shall deal with different, purely non-asymptotic aspects like unbiased estimation and invariance. In addition, we shall discuss systematically – apparently for the first time in literature – the multivariate and even the infinite-dimensional cases.

The contents are organized as follows. In Chapter 1 we introduce and investigate the L- and M-spaces of an experiment. It is shown why elements of the M-space are generalizations of random variables, and how one can define convex operations on, and multiplication of, those generalized random variables.

Chapters 2 and 3 deal with a method which yields a problem reduction, provided one knows that the relevant information is contained in a sub-σ-field $\mathcal{C}$ of $\mathcal{F}$. In the measure theoretic context, this reduction may be accomplished by means of the conditional expectation $E_P(\varphi | \mathcal{C})$ with respect to P of a random variable φ, given $\mathcal{C}$.

The counterparts of sub-σ-fields $\mathcal{C}$ of $\mathcal{F}$, namely sub-algebras $\mathcal{A}$ of the M-space $\mathcal{M}$, are treated in Chapter 2, where we shall also generalize the transition from φ to $E_P(\varphi | \mathcal{C})$. This yields a projection operator Π_P, the properties of which are well known in the context of one single probability space $(\Omega, \mathcal{F}, P)$. In this particular setting, some authors, e.g. [*Neveu* 1969, pp.149f.], use Π_P to define conditional expectations exactly.

In Chapter 3 we concentrate on the case where the reduction mentioned above is possible without loss of information: in this situation, the sub-algebra $\mathcal{A}$ is termed "sufficient" in analogy to the classical notion of sufficient sub-σ-fields $\mathcal{C}$. We shall derive some results useful for the treatment in the subsequent Chapters 4 and 5. In the course of this development, it will turn out that generalizing random variables yields results which in the classsical context hold only if the experiment is dominated.

While the results of Chapters 1, 2, and 3, seem to be of some interest for the general theories of statistical inference, in particular in those dealing with decision-theoretic questions, Chapters 4 and 5 focus on aspects of estimation theory. However, invariance and sufficiency are important concepts in the theory of testing hypotheses, cf. [*Lehmann 1959*]. Hence some of the results presented in Chapter 5 may be of some relevance for hypotheses testing, too.

Chapter 4 deals with optimality and admissibility in unbiased estimation. A parameter ϑ of interest is (unbiasedly) estimable, if there is a generalized random variable $f \in \mathcal{M}$ fulfilling

$$f(P) = \vartheta(P) \quad \text{for all } P \in \mathcal{P}.$$

In this case we call f a "generalized unbiased estimator" of ϑ. The risk of f with respect to a convex loss function is defined with the help of considerations in Chapter 1. If f minimizes this risk among all generalized unbiased estimators of ϑ, then we obviously call f "optimal" with respect to this loss function; if there is no alternative generalized unbiased estimator with lower risk, then f is termed "admissible".

These concepts are compared with each other and put into relation with sufficiency. Again, we shall derive statements on optimality and admissibility, which for undominated experiments are valid only in a restricted sense if one chooses the classical approach, and which can be modified in a straightforward way in order to hold for generalized estimators, without any assumptions on the structure of the experiment. One example for this situation is the criterion for the existence of nontrivial optimal unbiased estimators which we shall specify in section 4.5.

If we know that the models $(\Omega, \mathcal{F}, P)$, $P \in \mathcal{P}$, are left unchanged under some transformations, then this knowledge is a gain of information which is closely related to sufficiency, and therefore also connected with optimality and admissibility. This is the central aspect of Chapter 5, in which we deal with invariance with respect to groups G of transformations which act on the experiment. Again, we shall obtain results on generalized random variables that in the classical context can be derived only under additional assumptions, e.g. about the structure of the group G. However, we shall exploit further information about group structure, or about the way G acts on the experiment, to calculate explicitly the sufficient projection Π_G, which enables us to transform generalized estimators into invariant ones without losing relevant information. Furthermore, we shall proceed along the lines suggested by *Basu* [1970] by investigating maximal invariants under G in the metrically transitive case.

In a situation where several or even an infinity of observations can be obtained simultaneously, it has proved useful to consider random vectors, or random elements in infinite-dimensional linear spaces, respectively. Therefore at the end of each chapter we transfer the obtained results to generalized random elements which are defined analogously by functional analytic means. Thus we are able to cope with invariance, admissiblity, and optimality in the estimation of multivariate or even infinite dimensional parameters, too.

The Appendix is devoted to *Torgersen's* [1980] alternative way to generalize random variables. We shall show that both approaches – *LeCam's* and *Torgersen's* – yield structurally the same concepts. Furthermore we shall deal with a characterization of an important class of undominated experiments, and finally we shall indicate how one may extend the results presented in this book.

1. Generalized random elements and convex operations

In this chapter, we shall develop the tools for a functional analytic treatment of statistical experiments. After introducing the basic notions in section 1.1., we define in section 1.2. convex operations on, and the multiplication of, generalized random variables. Section 1.3. deals with the generalization of *Banach* space valued random elements, while section 1.4. is devoted to the study of a class of convex loss functions defined on a *Banach* space. A useful representation theorem for generalized random elements, which is stated and proved in section 1.5., closes this introductory chapter.

1.1. Experiments and their L- and M-spaces

Let us start by recalling a few essential facts about measures. We consider a measurable space $(\Omega, \mathcal{F})$: Ω is the set of (abstract) sample points – which may be realizations of a random variable $(\Omega = \mathbb{R})$, or of a random vector $(\Omega = \mathbb{R}^n)$, or even paths of a stochastic process (Ω being a suitably chosen, infinite-dimensional linear space) – and $\mathcal{F}$ is a σ-field in Ω representing the system of (composite) random events of interest. By $ca(\Omega, \mathcal{F})$ we denote the vector-lattice of all signed finite measures on $(\Omega, \mathcal{F})$, the ordering between two elements $\mu, \nu \in ca(\Omega, \mathcal{F})$ being defined by

$$\mu \leq \nu \quad \text{if and only if} \quad \mu(A) \leq \nu(A) \text{ for all } A \in \mathcal{F}.$$

With respect to this ordering, the lattice $ca(\Omega, \mathcal{F})$ is order complete. This means that for any subset $M = \{\mu_i : i \in I\}$ of $ca(\Omega, \mathcal{F})$ fulfilling $\mu_i \leq \nu$ for all $i \in I$ and some $\nu \in ca(\Omega, \mathcal{F})$, the least upper bound of M w.r.t. $\leq$ exists as a signed finite measure, which we denote by $\sup M$ or $\sup_{i \in I} \mu_i$. Similarly, if $\mu_i \geq \nu$ for all $i \in I$ and some $\nu \in ca(\Omega, \mathcal{F})$, the signed finite measure $\inf M = \inf_{i \in I} \mu_i = -\sup_{i \in I}(-\mu_i)$ is the largest lower bound of $M = \{\mu_i : i \in I\}$. The set M, or, equivalently, the index set I may be of arbitrary cardinality.

6

Now we introduce some helpful notation: first, let $|\mu| := \sup\{\mu, -\mu\} \in ca(\Omega, \mathcal{F})$ (recall that for $A \in \mathcal{F}$ we have the inequality $|\mu(A)| \leq |\mu|(A)$ which in general is strict, see, e.g. [Dunford/Schwartz 1964, p.161], who use $|\mu|$ with a different meaning and write $v(\mu, .)$ instead of $|\mu|$). Furthermore, $\mu \perp \nu$ is an abbreviation for the relation $\inf\{|\mu|, |\nu|\} = o$. For $M \subseteq ca(\Omega, \mathcal{F})$ let $M_+ := \{\mu \in M : \mu \geq o\}$, and $M^\perp := \{\nu \in ca(\Omega, \mathcal{F}) : \nu \perp \mu \text{ for all } \mu \in M\}$, denote the positive part of M, or the orthogonal complement of M in $ca(\Omega, \mathcal{F})$, respectively.

The variational norm given by $\|\mu\| := |\mu|(\Omega)$ renders $ca(\Omega, \mathcal{F})$ an order complete *Banach* lattice. More precisely, $ca(\Omega, \mathcal{F})$ is an abstract L-space which means that the norm $\|.\|$ is additive on $ca(\Omega, \mathcal{F})_+$: for any two measures $\mu, \nu \in ca(\Omega, \mathcal{F})_+$ we have

$$\|\mu + \nu\| = (\mu + \nu)(\Omega) = \|\mu\| + \|\nu\|$$

holds for all $\{\mu, \nu\} \subseteq ca(\Omega, \mathcal{F})_+$ [Dunford/Schwartz 1964, p.163], [Schaefer 1974, pp.112f.]. A linear sub-space $\mathcal{D} \subseteq ca(\Omega, \mathcal{F})$ is called a "band" , if $\mathcal{D}$ is "solid", which means

$$\mu \in \mathcal{D} \text{ whenever } \mu \in ca(\Omega, \mathcal{F}) \text{ and } |\mu| \leq |\nu| \text{ for some } \nu \in \mathcal{D},$$

and if $\mathcal{D}$ is "sup-closed": whenever $\mu_i \in \mathcal{D}$ for all $i \in I$, I being an index set of arbitrary cardinality, then

$$\mu := \sup_{i \in I} \mu_i \in \mathcal{D} \text{ provided } \sup_{i \in I} \mu_i \text{ exists in } ca(\Omega, \mathcal{F}).$$

Let $\mathcal{P} \subseteq ca(\Omega, \mathcal{F})$ be a set of probability measures. The triplet $\mathcal{E} := (\Omega, \mathcal{F}, \mathcal{P})$ is called "(statistical) experiment". We now give three equivalent definitions for the L-space $\mathcal{L}$ of an experiment $\mathcal{E}$. In the course of proving these equivalences – and also in the sequel – it is convenient to denote by $\Xi(M)$ the system of all finite sub-sets of an arbitrary set M.

Proposition 1.1 and Definition:

(a) Let $\mathcal{L}_1$ be the smallest band (with respect to $\subseteq$) in $ca(\Omega, \mathcal{F})$ containing $\mathcal{P}$;

(b) let $\mathcal{L}_2$ the $\|.\|$-closure of $\mathcal{L}'$, where
$$\mathcal{L}' := \{\mu \in ca(\Omega, \mathcal{F}) : |\mu| \leq \textstyle\sum_{i=1}^{n} \alpha_i P_i \text{ for some } P_i \in \mathcal{P}, \alpha_i \geq 0, 1 \leq i \leq n; n \in \mathbb{N}\};$$

(c) let $\mathcal{L}_3 := \{\mu \in ca(\Omega, \mathcal{F}) : \mu \perp \sigma \text{ for all } \sigma \in \mathcal{P}^\perp\}$.

Then all three $\mathcal{L}_i$, $i \in \{1, 2; 3\}$, are identical.

$$\mathcal{L} := \mathcal{L}_1 = \mathcal{L}_2 = \mathcal{L}_3 \quad \text{and its topological dual} \quad \mathcal{M} := \mathcal{L}^*$$

are called "L-space of $\mathcal{E}$", and "M-space of $\mathcal{E}$", respectively.

Proof: We show $\mathcal{L}_1 \subseteq \mathcal{L}_2 \subseteq \mathcal{L}_3 \subseteq \mathcal{L}_1$.

(i) $\mathcal{L}'$ is an ideal, i.e. a linear subspace of $ca(\Omega, \mathcal{F})$ that is solid. Hence also $\mathcal{L}_2$ is an ideal, and a fortiori a sub-lattice of $ca(\Omega, \mathcal{F})$ [*Schaefer* 1974, p.84]. Suppose now that $\mu_i \in \mathcal{L}_2, i \in I$, and that $\mu := \sup_{i \in I} \mu_i$ exists in $ca(\Omega, \mathcal{F})$. Then

$$\mu = \sup_{\xi \in \Xi(I)} \mu_\xi ,$$

where we have put $\mu_\xi := \sup_{i \in \xi} \mu_i$. Since $(\mu - \mu_\xi)_{\xi \in \Xi(I)}$ is a family that is directed with respect to $\geq$ and has infimum o, and since $ca(\Omega, \mathcal{F})$ is an abstract L-space, we conclude $\mu \in \mathcal{L}_2$ [*Schaefer* 1974, pp.89,113]. Therefore $\mathcal{L}_2$ is a band fulfilling $\mathcal{P} \subseteq \mathcal{L}' \subseteq \mathcal{L}_2$, thus $\mathcal{L}_1 \subseteq \mathcal{L}_2$.

(ii) If $\sigma \in ca(\Omega, \mathcal{F})$ fulfills $\sigma \perp P$ for all $P \in \mathcal{P}$ then for any $\mu \in \mathcal{L}'$ we have

$$\inf\{|\mu|, |\sigma|\} \leq \inf\{\sum_i \alpha_i P_i, \sigma\} = \sum_i \alpha_i \inf\{P_i, \sigma\} = o,$$

thus $\mu \perp \sigma$, whence $\mathcal{L}' \subseteq \mathcal{L}_3$ results. Continuity of the map $(\mu, \nu) \mapsto \inf\{\mu, \nu\}$ yields $\mathcal{L}_2 \subseteq \mathcal{L}_3$.

(iii) Since $\mathcal{L}_3$ is a sub-lattice and $\mathcal{L}_1$ is a sub-vector-lattice of $ca(\Omega, \mathcal{F})$, it suffices to show the inclusion $(\mathcal{L}_3)_+ \subseteq (\mathcal{L}_1)_+$. Now suppose that there were a $\mu \in (\mathcal{L}_3)_+ \setminus (\mathcal{L}_1)_+$; then this would entail [*Schaefer* 1974, p.62] $\mu = \varrho + \sigma$, where $\varrho \in (\mathcal{L}_1)_+$ and $\sigma \in ca(\Omega, \mathcal{F})_+ \setminus \{o\}$ fulfilled $\sigma \perp P$ for all $P \in \mathcal{P}$, yielding the contradiction

$$o = \inf\{\mu, \sigma\} = \inf\{\varrho + \sigma, \sigma\} \geq \sigma > o. \qquad \square$$

To each experiment $\mathcal{E}$ there is associated the system of null sets

$$\mathcal{N} := \{A \in \mathcal{F} : P(A) = 0 \text{ for all } P \in \mathcal{P}\}.$$

If there exists a $Q \in ca(\Omega, \mathcal{F})$ such that $\mathcal{N} = Q^{-1}(\{0\})$, then $\mathcal{E}$ is called "dominated" and we write $\mathcal{P} \cong Q$. In this case, the L-space $\mathcal{L}$ of $\mathcal{E}$ is identical to the set of all measures in $ca(\Omega, \mathcal{F})$ that are absolutely continuous with respect to Q. If $\mathcal{P} \cong Q$, then $\mathcal{L}$ is, as a *Banach* lattice, isomorphic to $L^1(Q)$ (see, e.g. [*Strasser* 1985, p.227f.], where also the inclusion $\mathcal{L}_1 \subseteq \mathcal{L}_3$ in Proposition 1.1 above is shown for general $\mathcal{E}$. Cf. [*Siebert* 1979]).

The situation for undominated experiments is different. Note that $\mathcal{L}$ itself, as defined in Proposition 1.1 above, is an abstract L-space, since it inherits all the structure of $ca(\Omega, \mathcal{F})$. A representation theorem of *Kakutani* [1941a] states that every abstract L-space is isomorphic to $L^1(m)$, m being a *Radon* measure on a locally compact topological space. But in general m is neither finite nor even σ-finite. Moreover, the L-space $\mathcal{L}$ of $\mathcal{E}$ is in general not isomorphic, as a *Banach* lattice, to an $L^1(\mu')$, $\mu' \in ca(\Omega', \mathcal{F}')$ being a finite measure on an arbitrary measurable space $(\Omega', \mathcal{F}')$:

Example: Let $\mathcal{P}$ be an uncountable set of mutually orthogonal probability measures on $(\Omega, \mathcal{F})$. Then $\mathcal{L}$ cannot be isomorphic to $L^1(\mu')$ as a *Banach* lattice. Indeed, otherwise there existed an element, say $\lambda \in \mathcal{L}_+$, corresponding to $1_{\Omega'}$. Because of

$$\int |\zeta - \inf\{\zeta, n 1_{\Omega'}\}|\, d\mu' \to 0 \quad \text{as } n \to \infty \quad \text{for all } \zeta \in L^1(\mu')\,, \tag{$*$}$$

we have $\|P - P_n\| \to 0$ as $n \to \infty$, where for $n \in \mathbb{N}$ we have put $P_n := \inf\{P, n\lambda\} \in \mathcal{L}$. By Proposition 1.1, $n\lambda \in \mathcal{L}$ can be approximated by measures $\mu_{n,k} \in \mathcal{L}'$, $k \in \mathbb{N}$, i.e. $\|\mu_{n,k} - n\lambda\| \to 0$ as $k \to \infty$. By definition of $\mathcal{L}'$, there are $\alpha_{i,k,n} \geq 0$ and $P_{i,k,n} \in \mathcal{P}$, $i \in \{1, \ldots, m_{k,n}\}$, such that

$$|\mu_{n,k}| \leq \sum_{i=1}^{m_{k,n}} \alpha_{i,k,n} P_{i,k,n} \text{ for all } k \in \mathbb{N}, \text{ for all } n \in \mathbb{N}\,.$$

Now choose a $P \in \mathcal{P} \setminus \{P_{i,k,n} : i \in \{1, \ldots, m_{k,n}\}, k \in \mathbb{N}, n \in \mathbb{N}\}$. Since

$$\inf\{P, \mu_{n,k}\} \leq \inf\{P, |\mu_{n,k}|\} \leq \inf\{P, \sum_{i=1}^{m_{k,n}} \alpha_{i,k,n} P_{i,k,n}\} = o \text{ for all } k \in \mathbb{N}\,,$$

we get, according to [*Schaefer* 1974, p.83], $P_n = \inf\{P, n\lambda\} = \lim_{k \to \infty} \inf\{P, \mu_{n,k}\} \leq o$ for all $n \in \mathbb{N}$ yielding the contradiction $P = \lim_{n \to \infty} P_n \leq o$. Hence the assertion follows. Let us add the remark that the existence of an element $\lambda \in \mathcal{L}_+$ fulfilling $(*)$ already implies that $\mathcal{L}$ is isomorphic to $L^1(\mu')$, μ' being a finite measure on $(\Omega', \mathcal{F}')_+$, where Ω' is a compact topological space and $\mathcal{F}'$ is the system of the *Borel* sets in Ω' [*Schaefer* 1974, pp.98,100,114]. ⋈

Now we furnish $\mathcal{M}$ by the order induced by its predual $\mathcal{L}$:

$$\text{for } \{f, g\} \subseteq \mathcal{M} \text{ we put } f \leq g\,, \text{ if } f(\mu) \leq g(\mu) \text{ for all } \mu \in \mathcal{L}_+\,.$$

Since $\mathcal{L}$ is an (order complete) abstract L-space, its topological dual $\mathcal{M}$ is an order complete abstract M-space [*Schaefer* 1974, p.121], i.e.

$$\| \sup\{f, g\}\| = \max\{\|f\|, \|g\|\} \text{ whenever } \{f, g\} \subseteq \mathcal{M}_+\,,$$

where $\mathcal{M}_+ := \{f \in M : f \geq o\}$ denotes the positive part of a sub-set $M \subseteq \mathcal{M}$ and $\|f\| := \sup\{f(\mu) : \mu \in \mathcal{L}, \|\mu\| \leq 1\}$ is the usual functional norm on $\mathcal{M}$. Recall that in $\mathcal{L}$ as well as in $\mathcal{M}$, order structure and norm topology are linked by the following implication:

$$\text{if } |a| \leq |b| \text{ then } \|a\| \leq \|b\|\,.$$

Some further notation is useful: for $a \in \mathcal{L}$ or $a \in \mathcal{M}$, we put $a_+ := \sup\{a, o\}$, $a_- := \sup\{-a, o\}$, $|a| := \sup\{a, -a\}$. We then have $a_- \perp a_+$, $a = a_+ - a_-$, and $|a| = a_+ + a_-$.

A particular element of $\mathcal{M}$ is given by the functional $e : \mu \mapsto \mu(\Omega)$ from $\mathcal{L}$ to $\mathbb{R}$. This element e is an order unit in $\mathcal{M}$, which means that for all $f \in \mathcal{M}$ there is an $\alpha \geq 0$ with $|f| \leq \alpha e$. Furthermore, the identity $\|f\| = \inf\{\alpha \geq 0 : |f| \leq \alpha e\}$ holds for all $f \in \mathcal{M}$ [*Vulikh* 1967, p.260].

In order to see why $\mathcal{M}$ generalizes the space of equivalence classes of bounded random variables, we introduce the space

$$X := \{\varphi : \Omega \to \mathbb{R} \ : \ \varphi \ \mathcal{F}\text{-measurable and bounded}\} \,.$$

On X we define the following equivalence relation $\sim$: for any pair $\{\varphi, \psi\} \subseteq X$,

$$\varphi \sim \psi \ \text{means that} \ \varphi = \psi \ \text{holds } P\text{-almost surely for all } P \in \mathcal{P} \,.$$

Now we shall show that there is an order-preserving, linear isomorphism between the factor $X/_\sim$ and the space $\dot{X} := \{\dot{\varphi} : \varphi \in X\} \subseteq \mathcal{M}$, where the functional $\dot{\varphi}$ is given by the map

$$\dot{\varphi} : \mu \mapsto \int \varphi \, d\mu$$
$$\mathcal{L} \to \mathbb{R}$$

for any function $\varphi \in X$ (with this notation we have, for instance $(\alpha 1_\Omega)^\cdot = \alpha e$ for all $\alpha \in \mathbb{R}$). The isomorphim between $X/_\sim$ and $\dot{X}$ is established by the following arguments: at first note that whenever $\varphi \sim \psi$, the set $N := \{\omega \in \Omega : \varphi(\omega) \neq \psi(\omega)\}$ belongs to $\mathcal{N}$, whence $\mu(N) = 0$ results for all $\mu \in \mathcal{L}$. This is an immediate consequence of the representation of $\mathcal{L}$ in Proposition 1.1(b). Furthermore, we shall now show that the supremum formed with respect to the ordering in $\mathcal{M}$ fulfills $\dot{\chi} = \sup\{\dot{\varphi}, \dot{\psi}\}$, where $\chi(\omega) := \max\{\varphi(\omega), \psi(\omega)\}$ defines a random variable $\chi \in X$, if φ and ψ belong to X. To this end define, for $\mu \in \mathcal{L}_+$, two measures ν and σ by $\nu(A) := \mu(A \cap \{\varphi > \psi\})$ and $\sigma(A) := \mu(A \cap \{\varphi \leq \psi\})$ for all $A \in \mathcal{F}$. By Proposition 1.1(b), the relations $|\nu| \leq \mu$ and $|\sigma| \leq \mu$ entail $\{\nu, \sigma\} \subseteq \mathcal{L}$, whence we deduce in turn

$$\sup\{\dot{\varphi}, \dot{\psi}\}(\mu) \leq \dot{\chi}(\mu) = \int_{\{\varphi > \psi\}} \varphi \, d\mu + \int_{\{\varphi \leq \psi\}} \psi \, d\mu = \dot{\varphi}(\nu) + \dot{\psi}(\sigma) \leq$$
$$\leq \sup\{\dot{\varphi}, \dot{\psi}\}(\nu) + \sup\{\dot{\varphi}, \dot{\psi}\}(\sigma) = \sup\{\dot{\varphi}, \dot{\psi}\}(\mu) \,.$$

Thus $X/_\sim$ is isomorphic to $\dot{X}$.

Now we shall show that $\mathcal{M}$ is in a certain sense the completion of $\dot{X}$: let $\sigma(\mathcal{M}, \mathcal{L})$ denote the weak-star topology on the dual $\mathcal{M}$ of $\mathcal{L}$: a net $(f_i)_{i \in I}$ is said to converge to an element $f \in \mathcal{M}$ w.r.t. $\sigma(\mathcal{M}, \mathcal{L})$ as $i \in I$, whenever $f_i(\mu) \to f(\mu)$ as $i \in I$ for all $\mu \in \mathcal{L}$. The following result proves that $\mathcal{M}$ coincides with the $\sigma(\mathcal{M}, \mathcal{L})$-closure of $\dot{X}$:

Proposition 1.2:

For every $f \in \mathcal{M}$ there is a net $(\varphi_i)_{i \in I}$ in X such that $|\varphi_i(\omega)| \leq \|f\|$ for all $\omega \in \Omega$ and all $i \in I$ as well as

$$\int \varphi_i d\mu \to f(\mu) \text{ as } i \in I \text{ for all } \mu \in \mathcal{L}.$$

If $f \in \mathcal{M}_+$ holds, we can even assume $\varphi_i \geq 0$ for all $i \in I$.

Proof: Lemma 41.9 in [*Strasser 1985*] yields the existence of a net $(\varrho_i, \psi_i)_{i \in I}$ in $X \times X$ fulfilling $0 \leq \varrho_i \leq \|f_+\|$ and $0 \leq \psi_i \leq \|f_-\|$ for all $i \in I$ as well as

$$\dot{\varrho}_i \to f_+ \text{ and } \dot{\psi}_i \to f_- \text{ w.r.t. } \sigma(\mathcal{M}, \mathcal{L}) \text{ as } i \in I.$$

For all $i \in I$, we have $\varphi_i := \varrho_i - \psi_i \in X$ and $-\|f_-\| \leq \varphi_i \leq \|f_+\|$. Furthermore φ_i clearly converges to $f_+ - f_- = f$ as $i \in I$ with respect to $\sigma(\mathcal{M}, \mathcal{L})$. Our claim now follows from the observation that $\{f_+, f_-\} \subseteq \mathcal{M}_+$ and $\sup\{f_+, f_-\} = |f|$ together entail the equality $\max\{\|f_+\|, \|f_-\|\} = \|\sup\{f_+, f_-\}\| = \|f\|$. $\qquad\square$

The considerations above justify the interpretation of the elements of $\mathcal{M}$ as generalized random variables. In the dominated case $\mathcal{P} \cong Q$, the space $\mathcal{M}$ clearly coincides with $\dot{X}$. Thus, every dominated experiment is also coherent in the sense that $\mathcal{M} = \dot{X}$ holds, the reverse implication being false in general [*Hasegawa/Perlman 1974, 1975; Pitcher 1965*]. Moreover, if $\mathcal{P} \cong Q$, then $X/\sim$ is the same as $L^\infty(Q)$. Hence the fact that $X/\sim$ and $\dot{X}$ are isomorphic reduces in the dominated case to the familiar isomorphism between $[L^1(Q)]^*$ and $L^\infty(Q)$.

Let us now turn to a closer study of the structure of $\mathcal{M}$. At first we note for easy reference in further use the following relationship between the orderings in $\mathcal{L}$ and in $\mathcal{M}$:

Proposition 1.3:
(a) For $\mu \in \mathcal{L}$,
$$\mu \in \mathcal{L}_+ \quad \text{if and only if} \quad f(\mu) \geq 0 \text{ for all } f \in \mathcal{M}_+.$$

(b) For $f \in \mathcal{M}_+$,
$$f = o \quad \text{if and only if} \quad f(P) = 0 \text{ for all } P \in \mathcal{P}.$$

Proof: (a) is trivial by $\dot{X} \subseteq \mathcal{M}$. For a more general statement ($\mathcal{L}$ a normed vector lattice) we refer to [*Vulikh* 1967, p.269].

(b): for $f \in \mathcal{M}_+$ fulfilling $f(\mathcal{P}) = \{0\}$ we have, by

$$|f(\mu)| \leq f(|\mu|) \leq \sum_{i=1}^{n} \alpha_i f(P_i) = 0$$

for all $\mu \in \mathcal{L}'$ also $f(\mathcal{L}') = \{0\}$, whence, by continuity, Proposition 1.1(b) yields $f = o$. Again, the statement remains true under far more general circumstances: if we replace, for instance, the space $\mathcal{M}$ with the set $\mathcal{L}_{00}^*$ of all order continuous linear functionals on $\mathcal{L}$ [*Schaefer* 1974, p.71], then for any $f \in (\mathcal{L}_{00}^*)_+$ the set $\tilde{\mathcal{L}}_f := \{\mu \in \mathcal{L} : f(|\mu|) = 0\}$ is a band containing $\mathcal{P}$ [*Schaefer* 1974, p.78]. Proposition 1.1(a) and the fact that $\mathcal{L}^* = \mathcal{L}_{00}^*$ holds for all abstract L-spaces $\mathcal{L}$ [*Schaefer* 1974, p.113] now yield the desired result. $\qquad\square$

As we have seen in the example after Proposition 1.1, in general one cannot find a finite measure space $(\Omega', \mathcal{F}', \mu')$ such that $\mathcal{L} \simeq L^1(\mu'), \mu' \in ca(\Omega', \mathcal{F}')_+$ holds. Hence we have no representation of $\mathcal{M}$ as $L^\infty(\mu')$. But $\mathcal{M}$ is an order complete abstract M-space. Thus, according to the representation theorems of *Krein/Krein* [1940] and *Kakutani* [1941b], the space $\mathcal{M}$ is, as a *Banach* lattice, isometrically isomorphic to the space $C(Z)$ of all continuous, real-valued functions on a compact topological space Z. As usual, $C(Z)$ is equipped with the pointwise vector lattice operations and by the supremum norm. In other words, if $\hat{f} \in C(Z)$ corresponds to $f \in \mathcal{M}$ via this isomorphism, we have $\|f\| = \sup_{z \in Z} |\hat{f}(z)|$ and $\hat{e} = 1_Z$, as well as

$$\widehat{\sup\{f, g\}}(z) = \max\{\hat{f}(z), \hat{g}(z)\} \text{ for all } z \in Z ,$$

if f and g belong to $\mathcal{M}$. Since $\mathcal{M}$ is order complete, the isomorphism yields an additional topological property of the space Z: Z is "extremally disconnected". This means that the closure $\overline{U}$ of every open set $U \subseteq Z$ is open (and of course closed; a set that is both closed and open is said to be "clopen").

As a final observation in this section, we see that every $\mu \in \mathcal{L}$ corresponds to a *Radon* measure $\hat{\mu} \in M(Z) = [C(Z)]^*$ via the isomorphism between $\mathcal{M}$ and $C(Z)$. This follows from the canonical embedding of $\mathcal{L}$ in its topological bidual $\mathcal{L}^{**} = \mathcal{M}^*$ by evaluation. On the other hand there are, in general, *Radon* measures in $M(Z)$ that have no counterpart in $\mathcal{L}$, because reflexive abstract L-spaces already are finite dimensional [*Schaefer* 1974, p.128].

1.2. Convex operations and multiplication

Let $L : \mathbb{R} \to \mathbb{R}$ be a convex function. In order to define the risk of a generalized random variable w.r.t. a loss function given by L, we must extend the map $\dot\varphi \mapsto (L \circ \varphi)\dot{}$ defined on $\dot X$ to all of $\mathcal{M}$. Clearly, the functional $L \circ f : \mathcal{M} \to \mathbb{R}$ is not linear for $f \in \mathcal{M}$, if L is not linear; hence $L \circ f \notin \mathcal{M}$. It neither makes sense to consider general convex transformations $R : \mathcal{M} \to \mathcal{M}$, as the following example shows:

Example: Let $\Omega := [0,1]$ and $\mathcal{F}$ be the system of *Borel* sets in Ω. Denote by λ *Lebesgue* measure on $\mathcal{F}$, and $\mathcal{P} = \{\lambda\}$ as well as $\kappa := 2(1_\Omega + 1_{[0,\frac{1}{2}]})$. The convex map $r : \varphi \mapsto \varphi^\kappa$ gives rise to a convex transformation $R : \dot\varphi \mapsto [r(\varphi)]\dot{}$ on $\mathcal{M} \simeq L^\infty(\lambda)$, but for $\varphi = 2(1_\Omega + 1_{[\frac{1}{2},1]}) \in X$ we have

$$(R\dot\varphi)(\lambda) = \int r(\varphi)\, d\lambda = 16 < 81 = r([\int \varphi\, d\lambda]\, 1_\Omega)(\omega)$$

for all $\omega \in [0, \frac{1}{2}]$. Hence the essential inequality of *Jensen* is violated (see Theorem 2.7 and the subsequent example in section 2.3.) ⑅

Thus it seems reasonable to study a different transformation on $\mathcal{M}$, say $f \mapsto L * f$, which is based upon a convex function $L : \mathbb{R} \to \mathbb{R}$. Note that for $\hat f \in C(Z)$ we have $L \circ \hat f \in C(Z)$. Hence $L * f \in \mathcal{M}$ is well defined by the equation $\widehat{(L*f)} = L \circ \hat f$. However, in order to show that $f \mapsto L * f$, generalizes the operation $\varphi \mapsto L \circ \varphi$ on X, we need an internal representation of $L * f$ as an element of the abstract M-space $\mathcal{M}$. This result has the additional benefit in that it will enable us to calculate the "generalized risk" $(L*f)(P)$ for $P \in \mathcal{P}$, in terms of evaluation of L at the numbers $f(\overline\mu)$. Here and in the sequel, we denote for $\mu \in \mathcal{L}_+$ with $\mu \neq o$ by $\overline\mu := \frac{1}{e(\mu)}\mu$ the corresponding probability measure.

Note beforehand that a function $L : \mathbb{R} \to \mathbb{R}$ is convex if and only if there are sequences of real numbers $(\alpha_n)_{n\in\mathbb{N}}$ and $(\beta_n)_{n\in\mathbb{N}}$ such that

$$L(t) = \sup_{n\in\mathbb{N}}[\alpha_n t + \beta_n] \quad \text{for all } t \in \mathbb{R}$$

(cf. Proposition 1.11 below). Now, if $f \in \mathcal{M}$ and $\alpha \geq 0$ fulfill $|f| \leq \alpha e$, then

$$\alpha_n f + \beta_n e \leq (\alpha_n \alpha + \beta_n)e \leq L(\alpha)e, \quad \text{if } \alpha_n \geq 0,$$

as well as

$$\alpha_n f + \beta_n e \leq (\alpha_n(-\alpha) + \beta_n)e \leq L(-\alpha)e, \quad \text{if } \alpha_n < 0.$$

Hence we observe that the least upper bound

$$\sup_{n\in\mathbb{N}}\,[\alpha_n f + \beta_n e] \in \mathcal{M}$$

belongs to $\mathcal{M}$.

Proposition 1.4:

Let $L(t) = \sup_{n\in\mathbb{N}}[\alpha_n t + \beta_n], t \in \mathbb{R}$, be a convex function. Then the following assertions hold:

(a) if $f \in \mathcal{M}$, then $L*f = \sup_{n\in\mathbb{N}}[\alpha_n f + \beta_n e]$;

(b) if $f, g \in \mathcal{M}$ and $0 \le \alpha \le 1$, then $L*(\alpha f + (1-\alpha)g) \le \alpha L*f + (1-\alpha)L*g$;

(c) if $f \in \mathcal{M}$ and $\mu \in \mathcal{L}_+$ with $\mu \ne o$, then $(L*f)(\mu) \ge e(\mu)L\big(f(\overline{\mu})\big)$;

(d) if f and μ are as in (c), then

$$(L*f)(\mu) = \sup\{\sum_{i=1}^{n}[\alpha_i f + \beta_i e](\mu_i) : \mu_i \in \mathcal{L}_+, \sum_{i=1}^{n}\mu_i = \mu\,;\, n \in \mathbb{N}\}$$

$$= \sup\{\sum_{j=1}^{m}e(\mu_j)L\big(f(\overline{\mu}_j)\big) : \mu_j \in \mathcal{L}_+ \setminus \{o\}, \sum_{j=1}^{m}\mu_j = \mu\,;\, m \in \mathbb{N}\}\,;$$

(e) if $\varphi \in X$, then $L \circ \varphi \in X$ and $(L \circ \varphi)^{\cdot} = L*\dot{\varphi}$.

Proof: (a) is implied by the relation $L\big(\hat{f}(z)\big) = \sup_{n\in\mathbb{N}}[\alpha_n \hat{f}(z) + \beta_n]$ which holds for all $z \in Z$. Assertion (b) follows from (a). The identity $(\alpha_n f + \beta_n e)(\mu) = e(\mu)[\alpha_n f(\overline{\mu}) + \beta_n]$ and (a) together yield (c). Assertion (a) also entails the first equality in (d), cf. [*Schaefer 1974, p.72*]. To show the second, asssume that $\sum_{j=1}^{m}\mu_{i_j} = \mu$, where $\mu_{i_j} \in \mathcal{L}_+ \setminus \{o\}, j \in \{1,\ldots,m\}$, whereas $\mu_i = o$ for all $i \in \{1,\ldots,n\} \setminus \{i_1,\ldots i_m\}$. Then

$$\sum_{i=1}^{n}[\alpha_i f + \beta_i e](\mu_i) = \sum_{j=1}^{m}[\alpha_{i_j} f(\overline{\mu}_{i_j}) + \beta_{i_j}]e(\mu_{i_j})$$

$$\le \sum_{j=1}^{m}L\big(f(\overline{\mu}_{i_j})\big)e(\mu_{i_j})$$

$$\le \sum_{j=1}^{m}(L*f)(\mu_{i_j}) = (L*f)(\mu),$$

14

the last inequality following from assertion (c). Hence also (d) is proven. It remains to show assertion (e): since the function L maps bounded sets into bounded sets, we obtain $L \circ \varphi \in X$. Define a function $\varphi_n \in X$ by

$$\varphi_n(\omega) := \max_{1 \leq k \leq n} [\alpha_k \varphi(\omega) + \beta_k] \leq L(\varphi(\omega)), \quad \omega \in \Omega.$$

Then $\dot{\varphi}_n = \sup_{1 \leq k \leq n}[\alpha_k \dot{\varphi} + \beta_k e] \leq \dot{\varphi}_{n+1}$ for all $n \in \mathbb{N}$. This implies, by the monotone convergence theorem,

$$
\begin{aligned}
(L * \dot{\varphi})(\mu) &= \big(\sup_{n \in \mathbb{N}} \dot{\varphi}_n\big)(\mu) \\
&\geq \sup_{n \in \mathbb{N}} [\dot{\varphi}_n(\mu)] \\
&= \int [\sup_{n \in \mathbb{N}} \varphi_n(\omega)] \, \mu(d\omega) \\
&= \int L(\varphi(\omega)) \, \mu(d\omega) = (L \circ \varphi)^{\cdot}(\mu)
\end{aligned}
$$

for all $\mu \in \mathcal{L}_+$, yielding $L * \dot{\varphi} \geq (L \circ \varphi)^{\cdot}$. On the other hand, the relation

$$(\alpha_n \dot{\varphi} + \beta_n e)(\mu) = \int (\alpha_n \varphi + \beta_n) \, d\mu \leq \int (L \circ \varphi) \, d\mu = (L \circ \varphi)^{\cdot}(\mu)$$

holds for all $n \in \mathbb{N}$ and all $\mu \in \mathcal{L}_+$, entailing $L * \dot{\varphi} \leq (L \circ \varphi)^{\cdot}$. Hence $L * \dot{\varphi} = (L \circ \varphi)^{\cdot}$. $\qquad \square$

Remark: The construction $L * f = \sup_{n \in \mathbb{N}}[\alpha_n f + \beta_n e]$ is possible in every countably order complete vector lattice $\mathcal{M}$ with order unit e. However, this generalization is not too relevant since every vector lattice $\mathcal{M}$ having the above properties already is isometrically isomorphic to $C(Y)$, Y being compact and σ-extremally disconnected. This means that the closure of every open F_σ–set in Y is clopen. Hence, as in the definition preceding Proposition 1.4, we could have put alternatively $\widehat{L * f} = L \circ \hat{f}$ for $\hat{f} \in C(Y)$ (see also [*Vulikh* 1967, pp.131, 138, and 181ff.]). These considerations show furthermore that the representation in Proposition 1.4(a) is independent of the choice of the sequences $(\alpha_n)_{n \in \mathbb{N}}, (\beta_n)_{n \in \mathbb{N}}$, provided that $L(t) = \sup_{n \in \mathbb{N}}[\alpha_n t + \beta_n]$ is maintained for all $t \in \mathbb{R}$. $\qquad \triangle$

Now we extend pointwise multiplication $\varphi \cdot \psi$ on X, defined by $(\varphi \cdot \psi)(\omega) = \varphi(\omega)\psi(\omega)$, all $\omega \in \Omega$, in a way consistent with order and linear structure on $\mathcal{M}$ (cf.[*LeCam* 1964, Propositions 3 and 4]). To this end, we shall employ the convex function

$$S(t) := t^2 = \sup_{\alpha \in \mathbb{Q}}[2\alpha t - \alpha^2], \ t \in \mathbb{R}.$$

Proposition 1.5 and Definition:

(a) There is one and only one binary operation

$$(f, g) \mapsto f \cdot g,$$
$$\mathcal{M} \times \mathcal{M} \to \mathcal{M}$$

satisfying the following conditions:

(i) $f \cdot g \in \mathcal{M}_+$, if $\{f, g\} \subseteq \mathcal{M}_+$.

(ii) $f \cdot e = e \cdot f = f$ for all $f \in \mathcal{M}$.

(iii) $(f + g) \cdot h = f \cdot h + g \cdot h$ and $f \cdot (g + h) = f \cdot g + f \cdot h$, if $\{f, g, h\} \subseteq \mathcal{M}$.

This operation is related by $\widehat{f \cdot g} = \hat{f} \cdot \hat{g}$ to pointwise multiplication on $C(Z)$.

(b) $(\mathcal{M}, +, \cdot)$ is a commutative (and associative) *Banach* algebra over $\mathbb{R}$ with unit e. The map

$$f \mapsto f \cdot g$$
$$\mathcal{M} \to \mathcal{M}$$

is $\sigma(\mathcal{M}, \mathcal{L})$-$\sigma(\mathcal{M}, \mathcal{L})$-continuous for all $g \in \mathcal{M}$.

(c) The square function S has the following properties:

(i) $S * f = f \cdot f$ for all $f \in \mathcal{M}$;

(ii) $S * f = o$ if and only if $f = o$;

(iii) $S * (f + g) = S * f + S * g + 2 f \cdot g$ for all $\{f, g\} \subseteq \mathcal{M}$.

(d) For $\{\varphi, \psi\} \subseteq X$ we have $(\varphi \cdot \psi)^{\cdot} = \dot{\varphi} \cdot \dot{\psi}$.

Proof: (a) is proven, e.g. in [*Vulikh* 1967, p.149].

(b): first note that

$$|f \cdot g|^{\widehat{}} = |(\widehat{f \cdot g})| = |\hat{f} \cdot \hat{g}| = |\hat{f}| \cdot |\hat{g}| = |\widehat{|f| \cdot |g|}| \, .$$

Hence (a) yields $|f \cdot g| = |f| \cdot |g|$. Therefore, if $\|f\| \leq 1$ and $\|g\| \leq 1$, we have $|f| \leq e$, $|g| \leq e$, and, again by (a),

$$|f \cdot g| = |f| \cdot |g| \leq e \cdot e = e \, .$$

Thus $\|f \cdot g\| \leq 1$ holds. Because of $\|e\| = \sup_{z \in Z} |1_Z(z)| = 1$, the element e is a unit in the *Banach* algebra $\mathcal{M}$. It remains to show the stated continuity. To this end we define, for $\mu \in \mathcal{L}$ and $g \in \mathcal{M}$, a measure $\mu' \in M(Z)$ by

$$\mu'(U) := \int_U \hat{g} \, d\hat{\mu} \, , \quad U \text{ a Borel set in } Z \, .$$

16

Obviously, this measure fulfills the relation $|\mu'| \leq \|g\| |\hat{\mu}|$. Now $\mathcal{L}$ is, considered under the canonical embedding, a band in $\mathcal{L}^{**}$ [*Schaefer 1974*, p.113], whence there exists a measure $\nu \in \mathcal{L}$ with $\hat{\nu} = \mu'$. Thus we have for all $f \in \mathcal{M}$

$$(f \cdot g)(\mu) = \int \hat{f} \cdot \hat{g} \, d\hat{\mu} = \int \hat{f} \, d\mu' = f(\nu).$$

Now $\sigma(\mathcal{M}, \mathcal{L})$-$\sigma(\mathcal{M}, \mathcal{L})$-continuity of $f \mapsto (f \cdot g)$ follows (cf.[*Semadeni 1971*, p.486]).
(c) is clear by $\widehat{S * f} = S \circ \hat{f} = \hat{f} \cdot \hat{f}$.
(d) follows from (c) and Proposition 1.4(e). $\qquad\square$

Remark: The "internal representation" of $f \cdot g$ is thus given by

$$f \cdot g = \frac{1}{2}[S*(f+g) - S*f - S*g],$$

where $S*f = \sup_{\alpha \in \mathbb{Q}}[2\alpha f - \alpha^2 e]$. $\qquad\triangle$

Corollary 1.6 and Definition:

There is exactly one operation

$$(g, \mu) \mapsto g \cdot \mu,$$
$$\mathcal{M} \times \mathcal{L} \to \mathcal{L}$$

having the property $f(g \cdot \mu) = (f \cdot g)(\mu)$ for all $f \in \mathcal{M}$.
This operation is bilinear on $\mathcal{M} \times \mathcal{L}$ and fulfills $g \cdot \mu \geq o$ whenever $g \geq o$ and $\mu \geq o$.

Proof: Since by Proposition 1.5(b), the map $f \mapsto (f \cdot g)(\mu)$ is a $\sigma(\mathcal{M}, \mathcal{L})$-continuous linear functional on $\mathcal{M}$, it is given by the evaluation at a suitably chosen measure $\nu \in \mathcal{L}$ which we call $g \cdot \mu$. Because $\nu = \sigma$ whenever $f(\nu) = f(\sigma)$ holds for all $f \in \mathcal{M}$, the measure $g \cdot \mu$ is uniquely determined by g and μ. Linearity in g is a consequence of Proposition 1.5(a), whereas linearity in μ is evident. If $g \geq o$ and $\mu \geq o$, Proposition 1.5(a)(i) entails $f(g \cdot \mu) = (f \cdot g)(\mu) \geq 0$ for all $f \in \mathcal{M}_+$, which in turn implies $g \cdot \mu \in \mathcal{L}_+$ by Proposition 1.3(a). $\qquad\square$

Remark: The proof of Proposition 1.5(b) shows that

$$(\widehat{g \cdot \mu})(U) = \int_U \hat{g} \, d\hat{\mu}$$

holds for every *Borel* set U in Z. For $\varphi \in X$ and $\mu \in \mathcal{L}$ we have

$$(\dot{\varphi} \cdot \mu)(A) = 1_A(\dot{\varphi} \cdot \mu) = (1_A \cdot \dot{\varphi})(\mu) = (1_A \cdot \varphi)\dot{\,}(\mu) = \int_A \varphi \, d\mu \quad \text{for all } A \in \mathcal{F},$$

because of Proposition 1.4(d). $\qquad\triangle$

For convenience of the reader, we close this section by specifying the following result by *Alaoglu/Bourbaki* (see, e.g. [*Strasser 1985*, p.472]):

Theorem 1.7:

For every net $(W_i)_{i \in I}$ of bilinear maps from $\mathcal{M} \times \mathcal{L}$ to $\mathbb{R}$ with

$$|W_i(f, \mu)| \leq K\|f\| \|\mu\| \quad \text{for all } i \in I \text{ and all } (f, \mu) \in \mathcal{M} \times \mathcal{L},$$

there is a sub-net which converges w.r.t. the product topology in $\mathbb{R}^{\mathcal{M} \times \mathcal{L}}$ (i.e. point-wise) towards a bilinear map $W : \mathcal{M} \times \mathcal{L} \to \mathbb{R}$ fulfilling

$$|W(f, \mu)| \leq K\|f\| \|\mu\| \quad \text{for all } (f, \mu) \in \mathcal{M} \times \mathcal{L} .$$

Proof: By *Tychonow*'s theorem we know that the product space

$$\prod_{(f, \mu) \in \mathcal{M} \times \mathcal{L}} [-K\|f\| \|\mu\|, K\|f\| \|\mu\|]$$

is compact in $\mathbb{R}^{\mathcal{M} \times \mathcal{L}}$. $\qquad\square$

1.3. Generalized random elements taking values in a Banach space

Now we turn towards the generalization of random elements in *Banach* spaces. Let B be a real *Banach* space equipped with the norm $\|\|.\|\|$. We denote the (functional) norms of its topological dual B^*, and of its topological bidual B^{**}, also by $\|\|.\|\|$. For $(x, y) \in B \times B^*$ we put $\langle x, y \rangle := y(x)$ as well as $\langle \overline{x}, y \rangle := \overline{x}(y)$ for $(\overline{x}, y) \in B^{**} \times B^*$. In this way, we embed B canonically in B^{**}. Let X^B denote the space of all bounded maps $\Phi : \Omega \to B$ which are *Bochner* integrable with respect to P for all $P \in \mathcal{P}$. Again, for $\{\Phi, \Psi\} \subseteq X^B$ we write

$$\Phi \sim \Psi, \quad \text{if for all } P \in \mathcal{P} \text{ we have} \quad \Phi = \Psi \quad P\text{-almost surely.}$$

$\Phi \in X^B$ is clearly *Bochner* integrable with respect to all $\mu \in \mathcal{L}'$. If $\dot{\Phi} : \mathcal{L} \to B$ denotes the extension of the linear map $\dot{\Phi}' : \mathcal{L}' \to B$ given by $\dot{\Phi}'(\mu) := \int \Phi \, d\mu$, which is bounded by $\|\|\Phi\|\|_\infty := \sup_{\omega \in \Omega} \|\|\Phi(\omega)\|\|$, then, as in the one-dimensional case, the factor $X^B/{\sim}$ is isomorphic to the space $\dot{X}^B := \{\dot{\Phi} : \Phi \in X^B\}$. By

$$\mathcal{M}^B := \{F : \mathcal{L} \to B \text{ linear} : \|\|F\|\| := \sup_{\|\mu\| \leq 1} \|\|F(\mu)\|\| < +\infty\}$$

we denote the space of all continuous linear maps from $\mathcal{L}$ to B. Then

$$B \cdot e := \{x \cdot e : x \in B\} \subseteq \dot{X}^B \subseteq \mathcal{M}^B,$$

if $x \cdot e$ denotes the map from $\mathcal{L}$ to B given by $(x \cdot e)(\mu) = \mu(\Omega)x \in B$, all $\mu \in \mathcal{L}$.

Let $\mu \in \mathcal{L}$ and $y \in B^*$. If $F \in \mathcal{M}^B$, we write $\langle F, y \rangle(\mu)$ instead of $\langle F(\mu), y \rangle$, so that $\langle F, y \rangle \in \mathcal{M}$ holds true. Similarly, for $x \in B$ and $F \in \mathcal{M}^{B^*}$ we write $\langle x, F \rangle(\mu)$ instead of $\langle x, F(\mu) \rangle$, so that $\langle x, F \rangle \in \mathcal{M}$. For $\mathcal{D} \subseteq \mathcal{M}$, we introduce

$$\mathcal{D}^B := \{F \in \mathcal{M}^B : \langle F, y \rangle \in \mathcal{D} \quad \text{for all } y \in B^*\}$$

as well as

$$^B\mathcal{D} := \{F \in \mathcal{M}^{B^*} : \langle x, F \rangle \in \mathcal{D} \quad \text{for all } x \in B\}.$$

If $\mathcal{D} = \mathcal{M}$, we clearly have $^B\mathcal{M} = \mathcal{M}^{B^*}$; for shortness we therefore stick to the notation $^B\mathcal{M}$ instead of $\mathcal{M}^{B^*}$. The equality

$$\int \langle \Phi(\omega), y \rangle \mu(d\omega) = \langle \int \Phi(\omega)\mu(d\omega), y \rangle \quad \text{for all } \mu \in \mathcal{L}' \text{ and all } y \in B^*$$

yields, via Proposition 1.1(b), $\langle \dot{\Phi}, y \rangle = \langle \Phi, y \rangle^{\cdot}$. This follows by standard continuity arguments. Hence, for arbitrary $\mu \in \mathcal{L}$, the element $\dot{\Phi}(\mu) \in B$ is the *Pettis* integral of Φ with respect to μ. By consequence, we have $\dot{X}^B \subseteq (\dot{X})^B \subseteq \mathcal{M}^B$ (put $\mathcal{D} = \dot{X} \subseteq \mathcal{M}$).

Remark: The following inclusion is evident:

$$X^{B^*} \subseteq M^{\infty}_{B*}(B) := \{\Psi : \Omega \to B^* : \langle x, \Psi \rangle \in X \text{ for all } x \in B\}.$$

Furthermore, for $\Psi \in M^{\infty}_{B*}(B)$ the uniform boundedness principle yields $|||\Psi|||_\infty < +\infty$ (see, e.g. [*Hirzebruch/Scharlau* 1971, p.37]). If we put for $\mu \in \mathcal{L}$

$$\dot{\Psi}(\mu) : x \mapsto \int \langle x, \Psi(\omega) \rangle \, \mu(d\omega),$$
$$B \to \mathbb{R}$$

we see that $\dot{\Psi} \in {}^B\mathcal{M}$, hence $[M^{\infty}_{B*}(B)]^{\cdot} \subseteq {}^B\mathcal{M}$. The element $\int \Psi d\mu := \dot{\Psi}(\mu) \in B^*$ is the $\sigma(B^*, B)$-analogue of the *Pettis* integral of Ψ with respect to μ. $\qquad \triangle$

Similarly to the case $B = \mathbb{R}$ we can interpret an element $F \in \mathcal{M}^B$ – and, in an extended sense, also an element $F \in {}^{B^*}\mathcal{M} = \mathcal{M}^{B^{**}} \supseteq \mathcal{M}^B$ – as a generalized random element in B :

Proposition 1.8:

$\dot{X}^B$ is $\sigma(\mathcal{M},\mathcal{L})$-dense in $\mathcal{M}^B$ with respect to $\sigma(B,B^*)$ and $\sigma(\mathcal{M},\mathcal{L})$-dense in $^{B^*}\mathcal{M}$ with respect to $\sigma(B^{**},B^*)$; $\dot{X}^{B^*}$ is $\sigma(\mathcal{M},\mathcal{L})$-dense in $^B\mathcal{M}$ with respect to $\sigma(B^*,B)$. To be more specific,

(a) for every $F \in {}^{B^*}\mathcal{M}$ there is a net $(\Phi_i)_{i \in I}$ in X^B fulfilling

$$\langle \dot{\Phi}_i, y \rangle \to \langle F, y \rangle \quad \text{w.r.t.} \quad \sigma(\mathcal{M},\mathcal{L}) \text{ as } i \in I, \quad \text{for all } y \in B^*$$

and

$$|||\Phi_i|||_\infty \leq |||F||| \quad \text{for all } i \in I;$$

(b) for every $F \in {}^B\mathcal{M}$ there is a net $(\Psi_i)_{i \in I}$ in X^{B^*} fulfilling

$$\langle x, \dot{\Psi}_i \rangle \to \langle x, F \rangle \quad \text{w.r.t.} \quad \sigma(\mathcal{M},\mathcal{L}) \text{ as } i \in I, \quad \text{for all } x \in B$$

and

$$|||\Psi_i|||_\infty \leq |||F||| \quad \text{for all } i \in I.$$

Proof: (a) If the assertion were false, there were $F \in {}^{B^*}\mathcal{M}, \mu \in \mathcal{L}, \varepsilon > 0$, and $y \in B^*$, fulfilling $|||y||| = 1$ as well as

$$|\langle F, y \rangle(\mu) - \langle \dot{\Phi}, y \rangle(\mu)| \geq \varepsilon \quad \text{for all } \Phi \in X^B \text{ with } |||\dot{\Phi}||| \leq |||F|||\,.$$

In particular, one would obtain $|\langle F, y \rangle(\mu)| \geq \varepsilon$ (for $\Phi \equiv o$). By Proposition 1.2, this relation yielded the existence of a function $\varphi \in X$ fulfilling

$$\|\varphi\|_\infty \leq \|\langle F, y \rangle\| \leq |||F||| \quad \text{and} \quad |\langle F, y \rangle(\mu) - \dot{\varphi}(\mu)| < \frac{\varepsilon}{2},$$

which in turn entailed $|\dot{\varphi}(\mu)| \neq 0$. Now choose an $x \in B$ with $|||x||| = 1 \leq \langle x, y \rangle + \frac{\varepsilon}{2|\dot{\varphi}(\mu)|}$. By consequence, for $\Phi := x \cdot \varphi \in X^B$ we had

$$|||\Phi|||_\infty \leq |||x|||\,\|\varphi\|_\infty \leq |||F||| \quad \text{and} \quad |\langle \dot{\Phi}(\mu), y \rangle - \dot{\varphi}(\mu)| < \frac{\varepsilon}{2}$$

entailing a contradiction. By replacing $x \in B$ by $y \in B^*$ in the above considerations we obtain the proof of (b). $\qquad \square$

Proposition 1.9 and Definition:

Let $\Pi : \mathcal{M} \to \mathcal{M}$ be a linear map fulfilling $\|\Pi f\| \leq C\|f\|$ for all $f \in \mathcal{M}$.

(a) There exists one and only one linear map ${}^B\Pi : {}^B\mathcal{M} \to {}^B\mathcal{M}$ with

$$\langle x, {}^B\Pi F \rangle = \Pi \langle x, F \rangle \quad \text{for all } F \in {}^B\mathcal{M} \text{ and all } x \in B.$$

This map satisfies $|||{}^B\Pi F||| \leq C|||F|||$ for all $F \in {}^B\mathcal{M}$.

(b) If Π is $\sigma(\mathcal{M}, \mathcal{L})$-$\sigma(\mathcal{M}, \mathcal{L})$-continuous, there exists one and only one linear map $\Pi^B : \mathcal{M}^B \to \mathcal{M}^B$ fulfilling

$$\langle \Pi^B F, y \rangle = \Pi \langle F, y \rangle \quad \text{for all } F \in \mathcal{M}^B \text{ and all } y \in B^*.$$

This map also satisfies $|||\Pi^B F||| \leq C|||F|||$ for all $F \in \mathcal{M}^B$.

Proof: Uniqueness of the maps Π^B and ${}^B\Pi$ is clear.

(a): for fixed $(F, \mu) \in {}^B\mathcal{M} \times \mathcal{L}$, the functional $x \mapsto (\Pi\langle x, F \rangle)(\mu)$ from B to $\mathbb{R}$ is linear as well as $|||.|||$-continuous, and hence can be represented by an element $y \in B^*$ which we name by $({}^B\Pi F)(\mu) := y$. Evidently, the map $\mu \mapsto ({}^B\Pi F)(\mu)$ from $\mathcal{L}$ to B^* is linear and fulfills

$$|||({}^B\Pi F)(\mu)||| = \sup\{(\Pi\langle x, F \rangle)(\mu) : x \in B, |||x||| = 1\} \leq C|||F|||\,\|\mu\|$$

for all $\mu \in \mathcal{L}$. Hence we obtain ${}^B\Pi F \in {}^B\mathcal{M}$ as well as the inequality stated above.

(b): Fix $\mu \in \mathcal{L}$ and $F \in \mathcal{M}^B \subseteq \mathcal{M}^{B^{**}} = {}^{B^*}\mathcal{M}$. Since the map $y \mapsto \langle F, y \rangle$ from B^* to $\mathcal{M}$ is $\sigma(B^*, B)$-$\sigma(\mathcal{M}, \mathcal{L})$-continuous, $\sigma(\mathcal{M}, \mathcal{L})$-continuity of Π yields $\sigma(B^*, B)$-continuity of the map $y \mapsto \Pi\langle F, y \rangle(\mu) = \langle ({}^{B^*}\Pi F)(\mu), y \rangle$ from B^* to $\mathbb{R}$. Thus we have ${}^{B^*}\Pi F(\mu) = x \in B$, whence we deduce the stated properties of the map $\Pi^B := ({}^{B^*}\Pi)|_{\mathcal{M}^B} : \mathcal{M}^B \to \mathcal{M}^B$. $\quad\square$

Next we generalize the operation

$$(\varphi, \Phi) \mapsto \varphi \cdot \Phi$$
$$X \times X^B \to X^B$$

of pointwise scalar multiplication, defined by $(\varphi \cdot \Phi)(\omega) = \varphi(\omega)\Phi(\omega) \in B$, all $\omega \in \Omega$. For $f \in \mathcal{M}$, consider the map $\Pi_f : g \mapsto f \cdot g$ from $\mathcal{M}$ to itself, which fulfills the assumptions in Proposition 1.9(b). We are therefore able to give the following definition:

Proposition 1.10 and Definition:

Let $f \in \mathcal{M}$ and $F \in \mathcal{M}^B$. Then the operation given by $f \cdot F := (\Pi_f)^B F \in \mathcal{M}^B$ has the following properties:

(a) for all $y \in B^*$ and all $\mu \in \mathcal{L}$ we have

$$\langle f \cdot F, y \rangle = f \cdot \langle F, y \rangle \quad \text{and} \quad (f \cdot F)(\mu) = F(f \cdot \mu);$$

(b) the map
$$(f, F) \mapsto f \cdot F$$
$$\mathcal{M} \times \mathcal{M}^B \to \mathcal{M}^B$$

is bilinear;

(c) for $\varphi \in X$ and $\Phi \in X^B$ we have $\dot{\varphi} \cdot \dot{\Phi} = (\varphi \cdot \Phi)^{\cdot}$.

Proof: (a) and (b) are immediate consequences of Corollary 1.6.
(c) follows from (a) and from Proposition 1.5(d), since

$$\langle \dot{\varphi} \cdot \dot{\Phi}, y \rangle = [\dot{\varphi} \cdot \langle \Phi, y \rangle^{\cdot}] = (\varphi \cdot \langle \Phi, y \rangle)^{\cdot} = (\langle \varphi \cdot \Phi, y \rangle)^{\cdot} = \langle (\varphi \cdot \Phi)^{\cdot}, y \rangle$$

holds for all $y \in B^*$. $\qquad\qquad \square$

1.4. Convex loss functions on Banach spaces

In this section we introduce and investigate a class of convex functions on B and B^* for which the statements of Proposition 1.4 are valid even in arbitrary dimensions.

Proposition 1.11 and Definition:

A "CB function" L on B is a convex function $L : B \to \mathbb{R}$ which is bounded on bounded sets, i.e. for all $\rho > 0$ there is a C_ρ fulfilling $|L(x)| \leq C_\rho$ for all $x \in B$ with $|||x||| \leq \rho$. A "CB^* function" L on B^* is a $\sigma(B^*, B)$-lower semicontinuous CB function $L : B^* \to \mathbb{R}$. Then

(a) every CB function is $|||.|||$-continuous;

(b) every $|||.|||$-continuous convex $L : B \to \mathbb{R}$ has a representation

$$L(x) = \sup_{i \in I}[\langle x, y_i \rangle + \beta_i], \quad x \in B,$$

for a suitably chosen set $\{(y_i, \beta_i) : i \in I\} \subseteq B^* \times \mathbb{R}$;

(c) every convex, $\sigma(B^*, B)$-continuous function $L : B^* \to \mathbb{R}$ is a CB^* function;

(d) a function $L : B^* \to \mathbb{R}$ is convex and $\sigma(B^*, B)$-lower semicontinuous if and only if there is a representation

$$L(y) = \sup_{i \in I}[\langle x_i, y \rangle + \beta_i], \quad y \in B^*,$$

for a suitably chosen set $\{(x_i, \beta_i) : i \in I\} \subseteq B \times \mathbb{R}$;

(e) every CB function L is, with the help of the representation in (b), extendable to a CB^* function $\overline{L} : B^{**} \to \mathbb{R}$ by the formula

$$\overline{L}(x) = \sup_{i \in I}[\langle x, y_i \rangle + \beta_i], \quad x \in B^{**};$$

conversely, every CB^* function on B^{**} is an extension of a CB function on B;

(f) if B is separable, the index set I in (b) can be replaced by a countable sub-set.

Proof: (a): for $\{x,u\} \subseteq B$ fulfilling $0 < \max\{|||x|||, |||u|||\} \le \rho$ define $\alpha := \frac{|||x-u|||}{\rho + |||x-u|||}$
and $w := \frac{1}{\alpha}x - (\frac{1}{\alpha} - 1)u$. $|||w - x||| = \rho$ yields now $|||w||| \le 2\rho$ whence $x = \alpha w + (1 - \alpha)u$
implies

$$\frac{L(x) - L(u)}{|||x - u|||} \le \alpha \frac{L(w) - L(u)}{|||x - u|||} \le \frac{2C_{2\rho}}{\rho}$$

which proofs even *Lipschitz* continuity on the norm balls in B.

(b): the epigraph of L, that is the set

$$\mathrm{epi}L := \{(u,t) \in B \times \mathbb{R} : t \ge L(u)\}$$

is convex and closed in $B \times \mathbb{R}$ w.r.t. the product topology given by $|||.|||$ and $|.|$ on B and
$\mathbb{R}$, respectively. Let $s < L(u)$, i.e. $(u,s) \notin \mathrm{epi}L$. Then there is $(\overline{y}_u, \lambda_u) \in B^* \times \mathbb{R} \setminus \{(o,0)\}$
such that

$$\langle u, \overline{y}_u \rangle + \lambda_u s \le \langle x, \overline{y}_u \rangle + \lambda_u t \quad \text{for all } (x,t) \in \mathrm{epi}L.$$

In particular for $t = L(x)$ and $x = u$ it follows $\lambda_u \ge 0$. We even have $\lambda_u > 0$, since
otherwise $\langle ., \overline{y}_u \rangle$ would be bounded from below which implied the contradiction $\overline{y}_u = o$.
For $y_u := -\frac{1}{\lambda_u}\overline{y}_u \in B^*$ we thus deduce

$$\langle u, y_u \rangle - s \ge \langle x, y_u \rangle - L(x) \quad \text{for all } s < L(u),$$

which therefore is true even if $s = L(u)$. Hence for $\beta_u := L(u) - \langle u, y_u \rangle$ we have

$$L(x) \ge \sup_{u \in B}[\langle x, y_u \rangle + \beta_u] \quad \text{for all } x \in B.$$

Conversely we know of course that $L(x) = \langle x, y_x \rangle + \beta_x$ for all $x \in B$. This yields the
statement, putting $I = B$.

(c) is a consequence of *Alaoglu*'s theorem.

(d): every function of the specified form is, as pointwise supremum of a family of $\sigma(B^*, B)$-
continuous and convex functions, $\sigma(B^*, B)$-lower semicontinuous and convex. The converse
proof is an analogue to that of (b), the only differences being that it is appropriate to choose
$I = B^*$, and that the epigraph $\mathrm{epi}L$ is now closed in $B^* \times \mathbb{R}$ w.r.t. the product topology
given by $\sigma(B^*, B)$ and $|.|$ on B^* and $\mathbb{R}$, respectively.

(e): as a consequence of (b) and (d), it only remains to show that $\overline{L}$ is bounded on bounded
sets. Therefore assume that $x \in B^{**}$ with $|||x||| \le \rho$. Pick a net $(x_i)_{i \in I}$ in B fulfilling
$|||x_i||| \le \rho$ for all $i \in I$ and $x_i \to x$ w.r.t. $\sigma(B^{**}, B^*)$ as $i \in I$ (see, e.g. [*Dunford/Schwartz*
1964, p.424]). By (d) we have

$$\overline{L}(x) \le \limsup_{i \in I} L(x_i) \le C_\rho,$$

24

hence $\overline{L}$ is bounded from above by C_ρ on the ball $\{x \in B^{**} : |||x||| \leq \rho\}$. The equation $\frac{1}{2}x + \frac{1}{2}(-x) = o$ now entails

$$\overline{L}(x) \geq 2\overline{L}(o) - \overline{L}(-x) = 2L(o) - \overline{L}(-x) \geq -3C_\rho,$$

thus

$$|\overline{L}(x)| \leq 3C_\rho \quad \text{for all } x \in B^{**} \text{ with } |||x||| \leq \rho.$$

(f): pick a countable dense sub-set $\{u_n : n \in \mathbb{N}\}$ in B and define $y_n := y_{u_n}$ as well as $\beta_n := \beta_{u_n}$, where $\{(y_u, \beta_u) : u \in B\} \subseteq B^* \times \mathbb{R}$ is chosen as in (b). For arbitrary $u \in B$ fulfilling $|||u||| \leq \rho$ there is an $x \in B$ such that

$$|||x||| = 1 \quad \text{and} \quad \langle x, y_u \rangle \geq |||y_u||| - 1.$$

If we now put $w := u + \rho x$, we have $|||w||| \leq 2\rho$ which yields, by the proof of (a),

$$(|||y_u||| - 1)\rho \leq L(u) - L(w) \leq \frac{2C_{4\rho}}{2\rho}|||u - w||| = C_{4\rho},$$

and therefore entails the relation

$$|||y_u||| \leq C'_\rho := \frac{C_{4\rho}}{\rho} + 1.$$

For every $x \in B$ there is a sequence $(n_k)_{k \in \mathbb{N}}$ such that $|||x - u_{n_k}||| \to 0$ as $k \to \infty$. Now define

$$\rho := \sup_{k \in \mathbb{N}} |||u_{n_k}|||$$

and pick a $k \in \mathbb{N}$ fulfilling

$$L(x) \leq L(u_{n_k}) + \frac{\varepsilon}{2} \quad \text{and} \quad |||x - u_{n_k}||| \leq \min\{\rho, \frac{\varepsilon}{2C'_\rho}\} \, .$$

Then

$$\langle x, y_{n_k} \rangle + \beta_{n_k} \geq L(u_{n_k}) - |||x - u_{n_k}||| \, |||y_{n_k}||| \geq L(x) - \varepsilon$$

which implies

$$L(x) \leq \sup_{n \in \mathbb{N}} [\langle x, y_n \rangle + \beta_n] \, . \qquad \square$$

Remark: Most of the statements in Proposition 1.11 are well known, see, e.g. [*Laurent* 1972, p.340], [*Phelps* 1983, p.282] or Lemma 1 in [*Bomze* 1984]. If $B \simeq \mathbb{R}^n$ has finite dimension, every convex function on B is a CB (and a CB*) function. If $B = B^{**}$ is reflexive, the notions "CB function" and "CB* function" coincide because of Proposition 1.11 (a),(b),(d). In any case, by transition from B to $B^{**} \supseteq B$ and by extension of L to $\overline{L}$ we obtain in a canonical way CB* functions from CB functions. However, this procedure described in Proposition 1.11(e) is in general not one-to-one. In fact, a corresponding uniqueness property characterizes in the case of dual spaces $B = B'^*$ reflexivity of B' and thus reflexivity of B [*Dunford/Schwartz* 1964, p.67]: indeed, $B' \neq B'^{**} = B^*$ implies the existence of a pair $(\overline{x}, \overline{y}) \in B^{**} \times B^*$ fulfilling

$$\langle \overline{x}, y' \rangle = 0 \quad \text{for all } y' \in B' \quad \text{and} \quad \langle \overline{x}, \overline{y} \rangle = 1 \,,$$

since B' is $|||.|||$-closed in B^*. Now let

$$\{y_i : i \in I\} = \{y \in B^* : |||y||| = 1\} \quad \text{and} \quad \{y'_i : i \in I'\} = \{y' \in B' : |||y'||| = 1\} \,.$$

Then for all $x \in B$ we have

$$L_1(x) := \sup_{i \in I}\langle x, y_i \rangle = |||x||| \quad \text{and}$$
$$L_2(x) := \sup_{i \in I'}\langle x, y'_i \rangle = |||x||| \,,$$

while on the other hand we have

$$\overline{L_1}(\overline{x}) = |||\overline{x}||| > 0 = \overline{L_2}(\overline{x}) \,.$$

This observation is essentially due to *W. Schachermayer* (personal communication; cf. [*Schachermayer* 1976]) who specified the example

$$B' = c_0 \,, B = \ell^1 \,, B^* = \ell^\infty \simeq C(\beta\mathbb{N}) \,, B^{**} \simeq M(\beta\mathbb{N}) \,, \overline{x} = \delta_z, z \in \beta\mathbb{N} \setminus \mathbb{N} \,,$$

$\beta\mathbb{N}$ being the *Stone/Čech* compactification of $\mathbb{N}$. $\qquad\qquad\triangle$

Not every CB function on B^* is a CB* function; neither (a) nor (c) in Proposition 1.11 have a converse that is valid in general:

Example: Let $B = \ell^1$. For any element $x = (x_n)_{n \in \mathbb{N}} \in B^* = \ell^\infty$, put $L(x) := \limsup_{n \in \mathbb{N}} x_n$. This defines a CB function L on B^* which is not $\sigma(B^*, B)$-lower semi-continuous, since, e.g. for the sequence of elements $x_i := (-\frac{n}{n+i})_{n \in \mathbb{N}} \in B^*$ $(i \in \mathbb{N})$ we have, on one hand, $x_i \to o$ w.r.t. $\sigma(B^*, B)$ as $i \to \infty$, but, on the other hand, $L(x_i) = -1 < 0 = L(o)$. $\qquad\qquad\bowtie$

26

Example: For $B = \ell^2 = B^*$, the function

$$L(x) := |||x|||_2 = \sup_{y \in B^*; |||y|||=1} \langle x, y \rangle$$

is a CB^* function which is not $\sigma(B^*, B)$-continuous. On the other hand, the function

$$L(x) := \sum_{n=1}^{\infty} |x_n|^n , x = (x_n)_{n \in \mathbb{N}} \in B$$

is $|||.|||$-continuous and convex, but no CB function. $\bowtie$

By means of the preceding considerations we now are able to generalize the operation $f \mapsto L * f$ for the multidimensional case:

Proposition 1.12 and Definition:

Let $L : B \to \mathbb{R}$ be a CB function, and $F \in \mathcal{M}^B$. For $\{(y_i, \beta_i) : i \in I\} \subseteq B^* \times \mathbb{R}$ as in Proposition 1.11(b) we define

$$L * F := \sup_{i \in I}[\langle F, y_i \rangle + \beta_i e] .$$

Then the following assertions hold true:

(a) $L * F \in \mathcal{M}$ is well defined; if H is also in $\mathcal{M}^B$ and $0 \leq \alpha \leq 1$, then

$$L * (\alpha F + (1 - \alpha)H) \leq \alpha L * F + (1 - \alpha)L * H ;$$

(b) if $\mu \in \mathcal{L}_+$ with $\mu \neq o$, then $L * F(\mu) \geq e(\mu)L\big(F(\overline{\mu})\big)$;

(c) with μ as in (b), we have

$$L * F(\mu) = \sup\{\sum_{i \in \xi} e(\mu_i)L\big(F(\overline{\mu}_i)\big) : \mu_i \in \mathcal{L}_+ \setminus \{o\}, \sum_{i \in \xi} \mu_i = \mu; \xi \in \Xi(I)\}$$

$$= \sup\{\sum_{i \in \xi}[\langle F, y_i \rangle + \beta_i e](\mu_i) : \mu_i \in \mathcal{L}, \sum_{i \in \xi} \mu_i = \mu; \xi \in \Xi(I)\} ;$$

(d) the representation of $L * F$ is independent of the choice of $\{(y_i, \beta_i) : i \in I\}$, provided the equality in Proposition 1.11(b) can be maintained. In particular we have:

(i) for a CB^* function L on B^* and $\overline{F} \in {}^B\mathcal{M}$,

$$L * \overline{F} = \sup_{i \in I}[\langle x_i, \overline{F} \rangle + \beta_i e] ;$$

(ii) for $\overline{L}$ as in Proposition 1.11(e), and $\overline{F} \in {}^{B^*}\mathcal{M}$,

$$\overline{L} * \overline{F} = \sup_{i \in I}[\langle \overline{F}, y_i \rangle + \beta_i e] ,$$

and, in particular, $L * F = \overline{L} * \overline{F}$;

(e) if $\Phi \in X^B$, then $L \circ \Phi \in X$ and $(L \circ \Phi)^{\cdot} = L * \dot{\Phi}$.

Proof: Since

$$[\langle F, y_i \rangle + \beta_i e](\mu) = e(\mu)[\langle F(\overline{\mu}), y_i \rangle + \beta_i] \leq e(\mu)L\big(F(\overline{\mu})\big) \leq C_\rho e(\mu)$$

is true for $\rho := |||F|||$, assertion (b) and the first part of (a) follow. Convexity and assertion (c) are obtained as in Proposition 1.4. Assertion (d) is a consequence of (c). (e): the map Φ is $\mathcal{F}$–$|||.|||$-*Borel*-measurable and the function L is $|||.|||$-continuous. Hence $L \circ \Phi \in X$ results which in turn entails $(L \circ \Phi)^\cdot \in \mathcal{M}$. Since $\langle \Phi(\omega), y_i \rangle + \beta_i \leq L(\Phi(\omega))$ for all $\omega \in \Omega$ and all $i \in I$, we have

$$\langle \dot{\Phi}, y_i \rangle + \beta_i e \leq (L \circ \Phi)^\cdot \quad \text{for all } i \in I$$

and therefore

$$L * \dot{\Phi} \leq (L \circ \Phi)^\cdot. \tag{$*$}$$

If B is separable, Proposition 1.11(f) yields

$$L(x) = \sup_{n \in \mathbb{N}} [\langle x, y_n \rangle + \beta_n] \quad \text{for all } x \in B,$$

so that by the monotone convergence theorem we obtain for all $P \in \mathcal{P}$

$$(L \circ \Phi)^\cdot(P) = \sup_{n \in \mathbb{N}} \int \sup_{1 \leq k \leq n} [\langle \Phi, y_k \rangle + \beta_k]\, dP$$
$$\leq \sup_{n \in \mathbb{N}} [(\sup_{1 \leq k \leq n} [\langle \Phi, y_n \rangle + \beta_n e])(P)] = (L * \dot{\Phi})(P)$$

These considerations, together with $(*)$ and Proposition 1.3(a), now imply $L * \dot{\Phi} = (L \circ \Phi)^\cdot$. If B is not separable, we proceed as follows: let $P \in \mathcal{P}$ be fixed. Since Φ is *Bochner* integrable with respect to P, there exist (i) a separable, closed linear sub-space B' of B and (ii) a set $N_P \in \mathcal{F}$ fulfilling $P(N_P) = 0$ such that $\Phi'(\Omega) \subseteq B'$, if we put $\Phi' := 1_{N_P} \cdot \Phi$. Because B' is separable, $|||\Phi'||| \in X$ entails $\Phi' \in X^{B'}$. Now define $L' := L|_{B'}$. Then L' is a CB function on B', so that by the preceding arguments,

$$(L \circ \Phi)^\cdot(P) = (L' \circ \Phi')^\cdot(P) = (L' * (\dot{\Phi'}))(P)$$

results. As a consequence of $\dot{\Phi}(\overline{\mu}) = (\dot{\Phi'})(\overline{\mu}) \in B'$ for all $\mu \in \mathcal{L}_+ \setminus \{o\}$ fulfilling $o \leq \mu \leq P$, we finally obtain $(L * \dot{\Phi})(P) = (L' * (\dot{\Phi'}))(P)$ by using (c), whence, again with the help of relation $(*)$, the equality $L * \dot{\Phi} = (L \circ \Phi)^\cdot$ is shown also for general B. $\qquad\square$

Proposition 1.13 and Definition:

Let $L : \mathbb{R} \to \mathbb{R}$ be convex and $(x,y) \in B \times B^*$.

(a) By the equations
$$L^y(x) := L(\langle x, y \rangle) =: L^x(y),$$
a CB function L^y on B as well as a CB* function L^x on B^* are defined. In particular, the function $\overline{L}^y$ defined by $\overline{L}^y(x) = L(\langle x, y \rangle)$ from B^{**} to $\mathbb{R}$ is a CB* function on B^{**};

(b) For $F \in \mathcal{M}^B$ we have $L^y * F = \overline{L}^y * F = L * \langle F, y \rangle$. Similarly, for $F \in {}^B\mathcal{M}$ the equation $L^x * F = L * \langle x, F \rangle$ holds.

Proof: (a) is clear, (b) follows by Proposition 1.12(c) and Proposition 1.4(d). $\qquad\square$

1.5. Representation of generalized random elements

We close this chapter by proving a useful analogon to the representation theorems by Krein/Krein and Kakutani, adapted for ${}^B\mathcal{M}$:

Theorem 1.14:

Assume that the representation $\mathcal{M} \simeq C(Z)$ holds. Then the following assertions are true:

(a) there is a linear isometry (which we designate by $F \mapsto \hat{F}$) between ${}^B\mathcal{M}$ and $C^B(Z)$, the space of all $\sigma(B^*, B)$-continuous functions $\hat{F} : Z \to B^*$, equipped with the norm $|||\hat{F}|||_\infty := \sup_{z \in Z} |||\hat{F}(z)|||$. For all $x \in B$ and all $F \in {}^B\mathcal{M}$ we have
$$\widehat{\langle x, F \rangle} = \langle x, \hat{F} \rangle ;$$

(b) for a ÇB* function L on B^* and $F \in {}^B\mathcal{M}$ we have
$$L * F(\mu) = \int L \circ \hat{F} \, d\hat{\mu} \quad \text{for all } \mu \in \mathcal{L}.$$

Proof: At first consider a function $\hat{F} \in C^B(Z)$. Since, for $x \in B$, $z \mapsto \langle x, \hat{F}(z) \rangle$ is continuous and Z is compact, the principle of uniform boundedness yields $|||\hat{F}|||_\infty < +\infty$. Hence the function $L \circ \hat{F} : Z \to \mathbb{R}$ is also bounded as well as *Borel*-measurable which follows by lower semicontinuity. Thus the map

$$f : \mu \mapsto \int L \circ \hat{F} \, d\hat{\mu}$$

$$\mathcal{L} \to \mathbb{R}$$

is well defined, linear and bounded, which means $f \in \mathcal{M}$.

(a): for $(\hat{F}, x)$ as above let $f_x \in \mathcal{M}$ fulfill $\hat{f}_x = \langle x, \hat{F} \rangle$. Fix a measure $\mu \in \mathcal{L}$. Then the functional $x \mapsto f_x(\mu)$ from B^* to $\mathbb{R}$ is linear and continuous. Indeed we have

$$f_{\alpha x + x'}(\mu) = \int \hat{f}_{\alpha x + x'} \, d\hat{\mu} = \int \langle \alpha x + x', \hat{F} \rangle \, d\hat{\mu}$$

$$= \alpha \int \langle x, \hat{F} \rangle \, d\hat{\mu} + \int \langle x', \hat{F} \rangle \, d\hat{\mu} = \alpha f_x(\mu) + f_{x'}(\mu) \,,$$

as well as

$$|f_x(\mu)| \le |f_x|(|\mu|) = \int |\langle x, \hat{F} \rangle| \, d|\hat{\mu}|$$

$$\le \int |||\hat{F}(z)||| \; |||x||| \; |\hat{\mu}|(dz) \le |||\hat{F}|||_\infty |||x||| \; \|\mu\|.$$

Therefore there is exactly one $F(\mu) \in B^*$ fulfilling $f_x(\mu) = \langle x, F(\mu) \rangle$. Evidently the map $F : \mu \mapsto F(\mu)$ is linear and fulfills $|||F||| \le |||\hat{F}|||_\infty$.

Conversely, assume that $F \in {}^B\mathcal{M}$. Now the relation $\|\langle x, F \rangle\| \le |||F||| \; |||x|||$ yields $\langle x, F \rangle \in \mathcal{M}$, and thus $\widehat{\langle x, F \rangle} \in C(Z)$. Similarly, for fixed $z \in Z$ the functional $x \mapsto \widehat{\langle x, F \rangle}(z)$ from B to $\mathbb{R}$ is linear and $|||.|||$-continuous. Hence there is a unique $\hat{F}(z) \in B^*$ fulfilling

$$\widehat{\langle x, F \rangle}(z) = \langle x, \hat{F}(z) \rangle \quad \text{for all } x \in B \text{ and all } z \in Z.$$

$\sigma(B^*, B)$-continuity of $\hat{F}$ follows immediately by $\widehat{\langle x, F \rangle} \in C(Z)$ for all $x \in B$. Now choose for $z \in Z$ and arbitrary $\varepsilon > 0$ an element $x_z \in B$ with $|||x_z||| = 1$ fulfilling $|||\hat{F}(z)||| \le \langle x_z, \hat{F}(z) \rangle + \varepsilon$. By consequence,

$$|||\hat{F}(z)||| \le \widehat{\langle x_z, F \rangle}(z) + \varepsilon \le \|\widehat{\langle x_z, F \rangle}\|_\infty + \varepsilon$$

$$= \|\langle x_z, F \rangle\| + \varepsilon \le |||F||| \; |||x_z||| + \varepsilon = |||F||| + \varepsilon \,,$$

and this relation implies $|||\hat{F}|||_\infty \le |||F|||$.

(b): let $\{(x_i, \beta_i) : i \in I\} \subseteq B \times \mathbb{R}$ be as in Proposition 1.11(d). For $\mu \in \mathcal{L}_+$,

$$[\langle x_i, F \rangle + \beta_i e](\mu) = \int [\widehat{\langle x_i, F \rangle} + \beta_i] \, d\hat{\mu}$$

$$= \int [\langle x_i, \hat{F} \rangle + \beta_i] \, d\hat{\mu} \le f(\mu) \quad \text{for all } i \in I,$$

whence we get $L*F(\mu) \leq f(\mu)$. On the other hand, we have

$$f(\mu) = \int L \circ \hat{F}\, d\hat{\mu}$$

$$= \sup_{i \in I}[\langle x_i, \hat{F}(z)\rangle + \beta_i]\,\hat{\mu}(z)$$

$$\leq \int \widehat{L*F}(z)\,\hat{\mu}(dz)\,,$$

because (a) entails

$$\langle x_i, \hat{F}(z)\rangle + \beta_i = [\langle x_i, F\rangle + \beta_i e]\hat{\ } \leq \widehat{L*F}(z)$$

for all $z \in Z$ (and all $i \in I$). $\qquad\qquad\qquad\qquad\qquad\qquad\square$

Example: Let $\Omega = \mathrm{IN}, \mathcal{F} = 2^{\mathrm{IN}}$, and denote by $\mathcal{P}$ all probability measures in $ca(\Omega, \mathcal{F})$, so that $\mathcal{L} = ca(\Omega, \mathcal{F}) \simeq \ell^1$. Consequently, we have $\mathcal{M} \simeq \ell^\infty \simeq C(Z)$, where $Z = \beta\mathrm{IN}$ (cf. the remark after Proposition 1.11). Put $B = \ell^2$. For $z \in \mathrm{IN} \subseteq Z$, denote by $e_z \in \ell^2$ the sequence $e_z := \big(\delta_z(\{n\})\big)_{n \in \mathrm{IN}}$. By defining

$$\hat{F}(z) = \begin{cases} e_z & , z \in \mathrm{IN}, \\ o & , z \in Z \setminus \mathrm{IN}, \end{cases}$$

we obtain a function $\hat{F} \in C^B(Z)$ which is not $|||.|||$-continuous. For the CB (and CB*) function $L = |||.|||$ we therefore have $L \circ \hat{F} \notin C(Z)$. $\qquad\qquad\bowtie$

Remark: The preceding example shows that, by contrast to the case $B = \mathrm{IR}$, it is impossible to define $L*F$ via $\widehat{L*F} = L \circ \hat{F}$. Of course we could have proceeded alternatively by defining

$$L*F(\mu) := \int L \circ \hat{F}\, d\hat{\mu}\,,\ \mu \in \mathcal{L}\,. \qquad\qquad\qquad\triangle$$

Theorem 1.14 deals only with $^B\mathcal{M}$, instead of treating $\mathcal{M}^B$. That this is not a severe restriction, can be easily conceived by the following observation: if $L : B \to \mathrm{IR}$ is an arbitrary CB function, then B^{**} and an extension $\overline{L}$ of L on B^{**} satisfy the assumptions of Theorem 1.14. Hence $\langle \hat{F}, y\rangle = \widehat{\langle F, y\rangle}$ follows as well as $L*F(\mu) = \int \overline{L} \circ \hat{F}\, d\hat{\mu}$ for all $\mu \in \mathcal{L}$ and all $F \in \mathcal{M}^B \subseteq {}^{B^*}\!\mathcal{M}$, according to Proposition 1.12(d)(ii).

2. Sub-algebras and conditional expectations

In this chapter we deal with counterparts of sub-σ-fields $\mathcal{C}$ of $\mathcal{F}$, and that of conditional expectations with respect to $P \in \mathcal{P}$, given $\mathcal{C}$. It will turn out in section 2.1., that the corresponding concept is again a (continuous version of the) conditional expectation, given a σ-field of events rather in Z than in Ω, with respect to the measure $\hat{P} \in M(Z)$. These considerations enable us to propose in section 2.2. a (linear and positive) extension of the operation $\dot{\varphi} \mapsto [E_P(\varphi|\mathcal{C})]\dot{}$, $\varphi \in X$, on $\mathcal{M}$ which satisfies an analogue to the *Reynold* property (Proposition 2.5). Section 2.3. is devoted to appropriately generalized versions of *Jensen*'s inequality. Above and in the sequel, $E_P(\varphi|\mathcal{C})$ always denotes a version of the conditional expectation.

2.1. Representations of sub-algebras in $\mathcal{M}$

Let us begin with some technicalities: to each idempotent $s = s \cdot s \in \mathcal{M}$ there corresponds exactly one set $U \subseteq Z$ with $\hat{s} := 1_U$ and vice versa. Since $\hat{s} \in C(Z)$, the set U is clopen. We thus denote the idempotent element s with $\hat{s} = 1_U$ by s_U, and by $\mathcal{B}_M := \{U \subseteq Z : s_U \in M\}$ the system of clopen sub-sets in Z corresponding to a sub-set $M \subseteq \mathcal{M}$.

Proposition 2.1:

Let $e \in \mathcal{A} \subseteq \mathcal{M}$ and $\mathcal{A}$ be $\|.\|$-closed. Then the following assertions are equivalent:

(a) $\mathcal{A}$ is a sub-algebra of $\mathcal{M}$.

(b) $\mathcal{A}$ is a sub-vector lattice of $\mathcal{M}$.

(c) $\mathcal{B}_\mathcal{A}$ is an algebra of sets, and the linear hull of all idempotents in $\mathcal{A}$, i.e. the set $\{\sum_{i=1}^n \alpha_i s_{U_i} : U_i \in \mathcal{B}_\mathcal{A} , \alpha_i \in \mathbb{R}, 1 \leq i \leq n ; n \in \mathbb{N}\}$, is $\|.\|$-dense in $\mathcal{A}$.

If one (and therefore all) of the conditions in (a),(b),(c) are true, then

$$\sup_{i \in I} f_i \in \mathcal{A},$$

whenever $f_i \in \mathcal{A}$ fulfill $f_i \leq g \in \mathcal{M}$ for all $i \in I$, and if I is a countable index set. If $\mathcal{A}$ is in addition $\sigma(\mathcal{M}, \mathcal{L})$-closed, the above relation holds true for index sets I of arbitrary cardinality.

Proof: (a) $\Rightarrow$ (b): let p_n be polynomials over $\mathbb{R}$, $n \in \mathbb{N}$, such that

$$\sup_{|t| \leq \|f\|^2} |p_n(t) - \sqrt{t}| \to 0 \quad \text{as } n \to \infty.$$

From

$$\sup_{z \in Z} |p_n([\hat{f}(z)]^2) - |\hat{f}(z)|| \to 0 \quad \text{as } n \to \infty,$$

we deduce $|f| \in \mathcal{A}$ for all $f \in \mathcal{A}$, and thus $\mathcal{A}$ is a sub-vector lattice of $\mathcal{M}$.

(b) $\Rightarrow$ (c): evidently, for $\{U, U'\} \subseteq \mathcal{B}_\mathcal{A}$ we have

$$s_{U \cap U'} = \inf\{s_U, s_{U'}\} \in \mathcal{A}, \quad s_{U \cup U'} = \sup\{s_U, s_{U'}\} \in \mathcal{A}, \quad \text{and } s_{Z \backslash U} = e - s_U \in \mathcal{A}.$$

A $\|.\|$-closed sub-vector lattice $\mathcal{A}$ of $\mathcal{M}$ is itself an abstract M-space [*Schaefer 1974*, p.103]. $\mathcal{A}$ is even countably order complete: indeed, $f_n \in \mathcal{A}$ and $f_n \leq g$ for all $n \in \mathbb{N}$ entails $g_n := \sup\{f_1, \ldots, f_n\} \in \mathcal{A}$; furthermore, for $f := \sup_{n \in \mathbb{N}} g_n \in \mathcal{M}$ we have, by $g_n \leq g_{n+1}$ for all $n \in \mathbb{N}$ and *Dini's* theorem (on Z),

$$\|f - g_n\| \to 0 \quad \text{as } n \to \infty,$$

thus $\sup_{n \in \mathbb{N}} f_n = f \in \mathcal{A}$. If $\mathcal{A}$ is $\sigma(\mathcal{M}, \mathcal{L})$-closed, then the assumptions $f_i \in \mathcal{A}$ and $f_i \leq g$ for all $i \in I$ yield, by similar arguments, $\sup_{i \in I} f_i \in \mathcal{A}$: indeed, the net $(g_\xi)_{\xi \in \Xi(I)}$, with $g_\xi := \sup_{i \in \xi} f_i \in \mathcal{A}$, is directed with respect to set inclusion $\subseteq$, and converges to $\sup_{i \in I} f_i = \sup_{\xi \in \Xi(I)} g_\xi$ w.r.t. $\sigma(\mathcal{M}, \mathcal{L})$ as $\xi \in \Xi(I)$, since

$$\sup_{\xi \in \Xi(I)} (g_\xi(\mu)) = (\sup_{\xi \in \Xi(I)} g_\xi)(\mu)$$

holds for all $\mu \in \mathcal{L}_+$ [*Schaefer 1974*, p.72]. Hence $\mathcal{A}$ is isomorphic to $C(Y)$, where Y is a compact and σ-extremally disconnected topological space (see the remark following Proposition 1.4). If we denote this isomorphism again by $f \mapsto \hat{f}$, we obtain $\hat{s} = 1_U$ if $s = s \cdot s \in \mathcal{A}$, where $U \subseteq Y$ is clopen. Next we show that for any $\{u, v\} \subseteq Y$ with $u \neq v$

there is a clopen set $U \subseteq Y$ such that $u \in U$ and $v \notin U$. Then the assertion follows from the theorem of *Stone/Weierstraß*, since the above implies the existence of a function $\hat{f} \in C(Y)$ fulfilling $\hat{f}(u) = 0$ and $\hat{f}(v) = 1$. To this end, denote by $U = \overline{W}$ the closure of the open F_σ set

$$W := \{y \in Y : \hat{f}(y) < \tfrac{1}{2}\} = \bigcup_{n \in \mathbb{N}} \{y \in Y : \hat{f}(y) \leq \tfrac{1}{2} - \tfrac{1}{n}\}.$$

Then indeed $u \in U$ and $v \notin U$ holds, and U is clopen in Y.

(c) $\Rightarrow$ (a): $\mathcal{A}$ is evidently a linear sub-space of $\mathcal{M}$. For $\{U, U'\} \subseteq \mathcal{B}_\mathcal{A}$ we have $s_U \cdot s'_U = s_{U \cap U'} \in \mathcal{A}$, so that Proposition 1.5(b) yields the implication claimed. $\qquad \square$

As the following observation shows, $\mathcal{B}_\mathcal{A}$ is – "up to $\hat{\mathcal{L}}$-null sets" – identical with the sub-σ-field $\hat{\mathcal{A}}$ of *Baire* sets in Z which is generated by $\mathcal{B}_\mathcal{A}$. By consequence, this means that every $\hat{\mathcal{A}}$-measurable bounded function $r : Z \to \mathbb{R}$ has a continuous version $\hat{f}$ (where $f \in \mathcal{A}$) with respect to all measures $\hat{\mu}$ (where $\mu \in \mathcal{L}$), a fact that at first sight is somewhat surprising (cf. [*Dixmier* 1951]).

Theorem 2.2:

Let $\mathcal{A}$ be a $\|.\|$-closed sub-algebra and $e \in \mathcal{A}$. Then the following assertions hold:

(a) $\hat{f}$ is $\hat{\mathcal{A}}$-measurable for every $f \in \mathcal{A}$. In particular (putting $\mathcal{A} = \mathcal{M}$) $\hat{\mathcal{M}}$ is the system of all *Baire* sets in Z.

(b) If, conversely, a function $r : Z \to [0, 1]$ is $\hat{\mathcal{A}}$-measurable, then there is an element $f \in \mathcal{A}$ with $0 \leq f \leq e$ fulfilling

$$\hat{f}(z) = r(z) \quad \hat{\mu}\text{-almost surely for all } \mu \in \mathcal{L}.$$

Proof: (a) follows from Proposition 2.1(a),(c).
(b) The system

$$\mathcal{A}' := \{U \in \hat{\mathcal{A}} : \text{there is a set } U' \in \mathcal{B}_\mathcal{A} \text{ with } \hat{\mu}(U \triangle U') = 0 \text{ for all } \mu \in \mathcal{L}\}$$

is a monotone class: indeed, since $U \triangle U' = (Z \setminus U) \triangle (Z \setminus U')$, it is sufficient to show $U := \bigcup_{n \in \mathbb{N}} U_n \in \mathcal{A}'$ whenever $U_n \subseteq U_{n+1} \in \mathcal{A}'$ holds for all $n \in \mathbb{N}$. For all $n \in \mathbb{N}$ choose $U'_n \in \mathcal{B}_\mathcal{A}$ such that $\hat{\mu}(U_n \triangle U'_n) = 0$ for all $\mu \in \mathcal{L}$. Because all U'_n are open, the set $U' := \overline{\bigcup_{n \in \mathbb{N}} U_n}$ is clopen in Z. Thus we have $1_{U'} \in C(Z)$, whence $s_{U'} = \sup_{n \in \mathbb{N}} s_{U'_n} \in \mathcal{A}$

(cf. the proof of Proposition 2.1), yielding $U' \in \mathcal{B}_{\mathcal{A}}$. Now Proposition 2.1(c) shows that we may without loss of generality assume that $U'_n \subseteq U'_{n+1}$ for all $n \in \mathbb{N}$, which in turn is equivalent to $s_{U'_n} \leq s_{U'_{n+1}}$ for all $n \in \mathbb{N}$. Hence for $\mu \in \mathcal{L}_+$ it follows [*Schaefer* 1974, p.72]:

$$
\begin{aligned}
\hat{\mu}(U') = s_{U'}(\mu) &= (\sup_{n \in \mathbb{N}} s_{U'_n})(\mu) \\
&= \sup_{n \in \mathbb{N}} [s_{U'_n}(\mu)] = \sup_{n \in \mathbb{N}} \hat{\mu}(U'_n) \\
&= \hat{\mu}(\bigcup_{n \in \mathbb{N}} U'_n) = \hat{\mu}(\bigcup_{n \in \mathbb{N}} U'_n \cap U) + \hat{\mu}(\bigcup_{n \in \mathbb{N}} U'_n \setminus U) \\
&\leq \hat{\mu}(U' \cap U) + \sup_{n \in \mathbb{N}} \hat{\mu}(U'_n \setminus U) \leq \hat{\mu}(U' \cap U) + \sup_{n \in \mathbb{N}} \hat{\mu}(U'_n \setminus U_n) = \hat{\mu}(U' \cap U),
\end{aligned}
$$

thus $\hat{\mu}(U' \setminus U) = 0$. Conversely we have $\hat{\mu}(U \setminus U') = 0$ because of the easily obtainable inclusion $U \setminus U' \subseteq \bigcup_{n \in \mathbb{N}} (U_n \setminus U'_n)$. Hence $\hat{\mu}(U \triangle U') = 0$, yielding the desired relation $U \in \mathcal{A}'$. Since $\mathcal{A} \subseteq \mathcal{A}'$, we arrive via Proposition 2.1(c) and the monotone class theorem at $\hat{\mathcal{A}} = \mathcal{A}'$. If $r : Z \to [0,1]$ is $\mathcal{A}'$-measurable and $U_{in} \in \mathcal{A}', \alpha_{in} \geq 0$ fulfill

$$
r_n(z) := \sum_{i=1}^{k_n} \alpha_{in} 1_{A_{in}}(z) \uparrow r(z) \quad \text{as } n \to \infty \quad \text{for all } z \in Z ,
$$

then we choose $U'_{in} \in \mathcal{B}_{\mathcal{A}}$ such that $\hat{\mu}(U_{in} \triangle U'_{in}) = 0$ holds for all $\mu \in \mathcal{L}$, and put, for $n \in \mathbb{N}$,

$$
g_n := \inf\{\sum_{i=1}^{k_n} \alpha_{in} s_{U'_{in}}, e\} \in \mathcal{A} \quad \text{as well as} \quad f_n := \sup\{g_1, \ldots, g_n\} \in \mathcal{A} .
$$

Then we get $\hat{f}_n(z) = r_n(z)$ $\hat{\mu}$-almost surely for all $n \in \mathbb{N}$, and $f := \sup_{n \in \mathbb{N}} f_n \in \mathcal{A}$ fulfills $0 \leq f \leq e$. Furthermore, we have (cf. proof of Proposition 2.1) $\|f_n - f\| \to 0$ as $n \to \infty$. Therefore choosing for an arbitrary $U \in \hat{\mathcal{A}}$ a set $U' \in \mathcal{B}_{\mathcal{A}}$ with $\hat{\mu}(U \triangle U') = 0$ for all $\mu \in \mathcal{L}$, we see that for every measure $\mu \in \mathcal{L}_+$

$$
\begin{aligned}
\int_U \hat{f} \, d\hat{\mu} = \int_{U'} \hat{f} \, d\hat{\mu} &= \lim_{n \to \infty} \int_{U'} \hat{f}_n \, d\hat{\mu} \\
&= \lim_{n \to \infty} \int_{U'} r_n \, d\hat{\mu} = \int_{U'} r \, d\hat{\mu} = \int_U r \, d\hat{\mu}
\end{aligned}
$$

holds, which proves the assertion, since $\hat{f}$ is $\hat{\mathcal{A}}$-measurable according to (a). $\qquad\square$

Now we characterize versions of the conditional expectation, given $\hat{\mathcal{A}}$, which "belong to $\mathcal{A}$":

Proposition 2.3:

Let $\mathcal{A}$ be a $\|.\|$-closed sub-algebra containing e. If $f \in \mathcal{M}, h \in \mathcal{A}$, and $P \in \mathcal{P}$ are fixed, then the following assertions are equivalent:

(a) $(f \cdot g)(P) = (h \cdot g)(P)$ for all $g \in \mathcal{A}$;

(b) $\hat{h} = E_{\hat{P}}(\hat{f}|\hat{\mathcal{A}})$ $\hat{P}$-almost surely.

Proof: (a) $\Rightarrow$ (b) : Define

$$\mathcal{A}' := \{U \in \hat{\mathcal{A}} : \int_U \hat{h} \, d\hat{P} = \int_U \hat{f} \, d\hat{P}\}.$$

Because of (a), $\mathcal{B}_{\mathcal{A}} \subseteq \mathcal{A}'$. On the other hand, $\mathcal{A}'$ is a sub-σ-field of $\hat{\mathcal{A}}$, whence $\hat{\mathcal{A}} = \mathcal{A}'$ and therefore (b) follows.

(b) $\Rightarrow$ (a) is a consequence of $\mathcal{B}_{\mathcal{A}} \subseteq \hat{\mathcal{A}}$, Proposition 2.1(c), and Proposition 1.5(b). $\square$

For $f \in \mathcal{M}_+$ and $P \in \mathcal{P}$ we put

$$\mathcal{S}_{\mathcal{A}}(f, P) := \{h \in \mathcal{A}_+ : (f \cdot g)(P) = (h \cdot g)(P) \text{ for all } g \in \mathcal{A}\},$$

so that, by Theorem 2.2 and Proposition 2.3, $\mathcal{S}_{\mathcal{A}}(f, P) \neq \emptyset$. However, if there is a $U \in \mathcal{B}_{\mathcal{A}} \setminus \{\emptyset\}$ fulfilling $\hat{P}(U) = 0$, then we have $h + s_U \in \mathcal{S}_{\mathcal{A}}(f, P)$ for all $h \in \mathcal{S}_{\mathcal{A}}(f, P)$, hence in this case $\mathcal{S}_{\mathcal{A}}(f, P)$ contains more than one element.

Remark: Let $\mathcal{C}$ be a sub-σ-field of $\mathcal{F}$ and denote by $\mathcal{A}$ the $\sigma(\mathcal{M}, \mathcal{L})$-closure of $\dot{X}_{\mathcal{C}}$, where $X_{\mathcal{C}}$ is the system of all bounded, $\mathcal{C}$-measurable functions from Ω to $\mathbb{R}$ (thus $X_{\mathcal{C}} \subseteq X$). Then $\mathcal{A}$ is a $\sigma(\mathcal{M}, \mathcal{L})$-closed sub-algebra containing e (see [*Siebert* 1979]). Via Proposition 1.5(b), Proposition 1.2, and the observation that

$$(\dot{\varphi} \cdot \dot{\psi})(P) = ([E_P(\varphi|\mathcal{C})]^{\cdot} \cdot \dot{\psi})(P) \quad \text{holds for all } \psi \in X_{\mathcal{C}} ,$$

Proposition 2.3 yields

$$([E_P(\varphi|\mathcal{C})]^{\cdot})^{\hat{}}(z) = E_{\hat{P}}(\hat{\dot{\varphi}}|\hat{\mathcal{A}})(z) \quad \hat{P}\text{-almost surely,}$$

which in turn implies

$$[E_P(\varphi|\mathcal{C})]^{\cdot} \in \mathcal{S}_{\mathcal{A}}(\dot{\varphi}, P).$$

If $P(A) = 0$ for an $A \in \mathcal{C} \setminus \mathcal{N}$, then $s_U := \dot{1}_A \in \mathcal{A} \setminus \{o\}$, hence $U \in \mathcal{B}_{\mathcal{A}} \setminus \{\emptyset\}$ while $\hat{P}(U) = 0$. As emphasized by *LeCam* [1964] and by *Siebert* [1979], we thus can view $\sigma(\mathcal{M}, \mathcal{L})$-closed sub-algebras $\mathcal{A}$ in $\mathcal{M}$ containing e as adequate generalizations of sub-σ-fields $\mathcal{C}$ of $\mathcal{F}$, since suprema of arbitrary families in $\mathcal{A}$ again belong to $\mathcal{A}$. $\triangle$

2.2. Projections and conditional expectations

The following result shows how to choose exactly one element of $\mathcal{S}_{\mathcal{A}}(f, P)$ in an appropriate way to obtain a linear and positive map from $\mathcal{M}$ to $\mathcal{A}$, if $\mathcal{A}$ is $\sigma(\mathcal{M}, \mathcal{L})$-closed.

Theorem 2.4:

Let $P \in \mathcal{P}$ and $\mathcal{A}$ be a $\sigma(\mathcal{M}, \mathcal{L})$-closed sub-algebra of $\mathcal{M}$ containing e. Then there is a linear and positive map $\Pi_P : \mathcal{M} \to \mathcal{A}$ that fulfills

(a) $\widehat{\Pi_P f}(z) = E_{\hat{P}}(\hat{f} | \hat{\mathcal{A}})(z)$ $\hat{P}$-almost surely for all $f \in \mathcal{M}$;

(b) $\Pi_P g \leq g$ for all $g \in \mathcal{A}_+$;

(c) $\Pi_P \circ \Pi_P = \Pi_P$.

Proof: At first assume that $f \in \mathcal{M}_+$. By Proposition 1.5(b) $\mathcal{S}_{\mathcal{A}}(f, P)$ is $\sigma(\mathcal{M}, \mathcal{L})$-closed. Since for $\{h, k\} \subseteq \mathcal{S}_{\mathcal{A}}(f, P)$ we have

$$\inf\{\hat{h}(z), \hat{k}(z)\} = E_{\hat{P}}(\hat{f} | \hat{\mathcal{A}})(z) \quad \hat{P}\text{-almost surely,}$$

we conclude $\inf\{h, k\} \in \mathcal{S}_{\mathcal{A}}(f, P)$. The arguments used in the proof of Proposition 2.1 now yield

$$\Pi_P f := \inf \mathcal{S}_{\mathcal{A}}(f, P) \in \mathcal{S}_{\mathcal{A}}(f, P) \subseteq \mathcal{A}_+ \, .$$

Next we show that the map

$$f \mapsto \Pi_P f$$
$$\mathcal{A}_+ \to \mathcal{M}_+$$

is positively homogenous and additive: indeed, since $\alpha h \in \mathcal{S}_{\mathcal{A}}(\alpha f, P)$, if $\alpha \geq 0$ and if $h \in \mathcal{S}_{\mathcal{A}}(f, P)$, we have $\Pi_P(\alpha f) = \alpha \Pi_P f$; also, for $\{f, f'\} \subseteq \mathcal{M}_+$, we obtain

$$\overline{h} := \Pi_P(f + f') \leq \Pi_P f + \Pi_P f' \in \mathcal{S}_{\mathcal{A}}(f + f', P),$$

so that, putting

$$h := \inf\{\overline{h}, \Pi_P f\} \in \mathcal{A}_+ \quad \text{and} \quad h' := \overline{h} - h \in \mathcal{A}_+ \, ,$$

we get [cf. *Schaefer* 1974, p.53] $h \leq \Pi_P f$ and $h' \leq \Pi_P f'$. Hence for any $g \in \mathcal{A}_+$ we arrive, on one hand, at

$$(g \cdot h)(P) \leq (g \cdot \Pi_P f)(P) = (g \cdot f)(P) \quad \text{and}$$
$$(g \cdot h')(P) \leq (g \cdot \Pi_P f')(P) = (g \cdot f')(P),$$

whereas, on the other hand,

$$(g \cdot h)(P) + (g \cdot h')(P) = (g \cdot \overline{h})(P) = (g \cdot (f + f'))(P) = (g \cdot f)(P) + (g \cdot f')(P)\,,$$

so that

$$(g \cdot h)(P) = (g \cdot f)(P) \quad \text{and} \quad (g \cdot h')(P) = (g \cdot f')(P)\,,$$

implying $h \in \mathcal{S}_{\mathcal{A}}(f, P)$ as well as $h' \in \mathcal{S}_{\mathcal{A}}(f', P)$. This in turn entails $\Pi_P f = h$ and $\Pi_P f' = h'$, whence we finally conclude $\Pi_P f + \Pi_P f' = \Pi_P(f + f')$. Thus putting for arbitrary $f \in \mathcal{M}$

$$\Pi_P f := \Pi_P(f_+) - \Pi_P(f_-)\,,$$

we obtain a positive and linear map $\Pi_P : \mathcal{M} \to \mathcal{A}$ (see, e.g. [*Vulikh* 1967, p.205]) fulfilling evidently (a) and (b). To show that (c) holds, note that, on one hand, property (b) together with positivity of Π_P yields

$$\Pi_P(\Pi_P f) \leq \Pi_P f$$

for every $f \in \mathcal{M}_+$, and that, on the other hand, Proposition 2.3 and (a) entail

$$[\Pi_P(\Pi_P f) \cdot g](P) = [(\Pi_P f) \cdot g](P) = (f \cdot g)(P) \quad \text{for all } g \in \mathcal{A}.$$

These considerations show $\Pi_P(\Pi_P f) \in \mathcal{S}_{\mathcal{A}}(f, P)$, whence assertion (c) follows by construction of $\Pi_P f$. $\qquad\qquad\square$

That Π_P – even in an extended sense – shares the properties of a conditional expectation operator, is implied by the following observation which sharpens a result by *LeCam* [1964; 1986, p.59]:

Proposition 2.5:
 Let $\mathcal{A}$ be a $\|.\|$-closed sub-algebra in $\mathcal{M}$ containing e. Consider a positive and linear map $\Pi : \mathcal{M} \to \mathcal{A}$ fulfilling $\Pi g \leq g$ for all $g \in \mathcal{A}_+$. Then

$$\Pi(f \cdot g) = (\Pi f) \cdot g \quad \text{for all } f \in \mathcal{M} \text{ and all } g \in \mathcal{A}$$

as well as $\|\Pi f\| \leq \|f\|$ for all $f \in \mathcal{M}$.

38

Proof: Let $s = s \cdot s \in \mathcal{A}$, $f \in \mathcal{M}$. Then

$$|\Pi(s \cdot f)| \le \Pi|s \cdot f| = \Pi(s \cdot |f|) \le \Pi(s \cdot \|f\|e) \le \|f\|\Pi s \le \|f\|s$$

and also $|\Pi[(e - s) \cdot f]| \le \|f\| \cdot (e - s)$ results. Hence for

$$h := s \cdot \Pi f - \Pi(s \cdot f) = \Pi[(e - s) \cdot f] - (e - s)\Pi f$$

we arrive at

$$|h| \le s \cdot |\Pi f| + |\Pi(s \cdot f)| \le (\|f\|e + |f|) \cdot s \le 2\|f\|s$$

and as well obtain $|h| \le 2\|f\|(e - s)$. Proposition 1.5(a) now entails

$$S * h \le 4\|f\|^2 s \cdot (e - s) = o,$$

yielding via Proposition 1.5(c) the equality $h = o$, which in turn implies $\Pi(s \cdot f) = s \cdot \Pi f$. The first assertion now follows via Proposition 2.1(c) by linearity and continuity arguments. To show the second one, observe that the inequality $|f| \le \|f\|e$ entails via positivity and linearity of Π

$$|\Pi f| \le \Pi|f| \le \Pi(\|f\|e) = \|f\|\Pi e \le \|f\|e$$

and hence $\|\Pi f\| \le \|f\|\|e\| = \|f\|$. $\qquad\qquad\square$

Remark: *LeCam* [1964, p.1437] assumes – without specifying a proof – that a map $\Pi_P : \mathcal{M} \to \mathcal{A}$ exists fulfilling the assumptions as well as (a) and (b) in Theorem 2.4, which has the additional property $\Pi_P e = e$. According to Proposition 2.5 (putting $f = e$), this yields $\Pi_P g = g$ for all $g \in \mathcal{A}$ (cf. also [*LeCam* 1986, p.60] where this property is dropped again, but where $\sigma(\mathcal{M}, \mathcal{L})$-continuity of Π_P is stated). In the sequel we shall use neither of these properties. However, both $\Pi_P e = e$ and $\sigma(\mathcal{M}, \mathcal{L})$-continuity of Π_P (at least on $\|.\|$-balls in $\mathcal{M}$) pertain whenever P fulfills $\hat{P}(U) > 0$ for all $U \in \mathcal{B}_\mathcal{M} \setminus \{\emptyset\}$, because then we have $\mathcal{S}_\mathcal{A}(f, P) = \{\Pi_P f\}$ for all $f \in \mathcal{M}$ (to show $\sigma(\mathcal{M}, \mathcal{L})$-continuity, use Proposition 1.5(b) and *Alaoglu*'s theorem). $\qquad\qquad\triangle$

2.3. Jensen's inequality

Now we deal with the counterpart of *Jensen's* inequality: we treat the case described in the remark at the end of the preceding section separately, since a stronger result is obtainable here:

Proposition 2.6:

Let $\Pi : \mathcal{M} \to \mathcal{M}$ be a positive, linear map fulfilling $\Pi e = e$. Consider a CB function L on B and its extension $\overline{L}$ on B^{**} (cf. Proposition 1.11(e)). Then

$$\overline{L}*({}^{B^*}\Pi F) \leq \Pi(\overline{L}*F) \quad \text{for all } F \in {}^{B^*}\mathcal{M}$$

holds. If Π is $\sigma(\mathcal{M}, \mathcal{L})$-$\sigma(\mathcal{M}, \mathcal{L})$-continuous, we furthermore have

$$L*(\Pi^B F) \leq \Pi(L*F) \quad \text{for all } F \in \mathcal{M}^B \, .$$

Proof: Let $L(x) = \sup_{i \in I}[\langle x, y_i \rangle + \beta_i e]$, $x \in B$; then for all $i \in I$

$$\langle {}^{B^*}\Pi F, y_i \rangle + \beta_i e = \Pi[\langle F, y_i \rangle + \beta_i e] \leq \Pi(\overline{L}*F) \, .$$

The remaining assertion is derived from the equality

$$\Pi^B = {}^{B^*}\Pi|_{\mathcal{M}^B}$$

using Proposition 1.12(d). $\qquad\qquad\square$

Let us now treat the general case: the result stated in Theorem 2.7 below will suffice for the subsequent considerations, but it needs no additional assumptions concerning the map Π_P.

Theorem 2.7:

Let L be a CB* function on B^*, $P \in \mathcal{P}$, and $\Pi_P : \mathcal{M} \to \mathcal{M}$ be a map fulfilling the assertions as well as condition (a) in Theorem 2.4. Then for all $F \in {}^B\mathcal{M}$ we have

$$[L*({}^B(\Pi_P)F)](P) \leq [\Pi_P(L*F)](P) \, .$$

Proof: Consider an arbitrary decomposition of P into non-negative measures belonging to $\mathcal{L}$: $P = \sum_{j=1}^{n} \mu_j$, where $\mu_j \in \mathcal{L}_+$, and denote by

$$r_j := \frac{d\widehat{\mu_j}}{d\hat{P}} : Z \to [0,1]$$

arbitrary versions of the *Radon/Nikodym* densities with respect to $\hat{P} \in M(Z)$ of the measures $\widehat{\mu_j} \in M(Z)$, $1 \leq j \leq n$. If

$$L(y) = \sup_{i \in I}[\langle x_i, y \rangle + \beta_i] , \quad y \in B^* ,$$

then we obtain

$$
\begin{aligned}
[\langle x_i, {}^B(\Pi_P)F \rangle + \beta_i e](\mu_j) &= \int [\Pi_P \langle x_i, F \rangle + \beta_i e]\,\widehat{\ } \cdot r_j \, d\hat{P} \\
&= \int [E_{\hat{P}}(\langle \widehat{x_i, F} \rangle | \hat{\mathcal{A}}) + \beta_i] \cdot r_j \, d\hat{P} \\
&= \int E_{\hat{P}}(\langle x_i, \hat{F} \rangle + \beta_i | \hat{\mathcal{A}}) \cdot r_j \, d\hat{P} \\
&\leq \int E_{\hat{P}}(L \circ \hat{F} | \hat{\mathcal{A}}) \cdot r_j \, d\hat{P} \\
&= \int E_{\hat{P}}(\widehat{L * F} | \hat{\mathcal{A}}) \cdot r_j \, d\hat{P} \\
&= \int [\Pi_P(L*F)]\,\widehat{\ }\,\widehat{d\mu_j} = [\Pi_P(L*F)](\mu_j) .
\end{aligned}
$$

Proposition 1.12(c) yields the desired result. $\qquad\square$

Example: Let us continue the example at the beginning of section 1.2. Here we consider $B = \mathbb{R}$ and ask the question whether or not the convex operation $R : \dot{\varphi} \mapsto [r(\varphi)]\,\dot{}$ on $\mathcal{M}$ satifies an analogue to Theorem 2.7, namely $[R(\Pi_\lambda f)](\lambda) \leq [\Pi_\lambda(R(f))](\lambda)$. Recall that λ is in this example the only probability measure in $\mathcal{P}$. We choose the $\sigma(\mathcal{M}, \mathcal{L})$-closed sub-algebra $\mathcal{A} := \mathbb{R} \cdot e = \{\alpha e : \alpha \in \mathbb{R}\}$ and put $\Pi_\lambda f := f(\lambda)e$. As is easily seen, this map satisfies the conditions of Theorem 2.4. For $f = \dot{\varphi}$ with $\varphi = 2(1_\Omega + 1_{[\frac{1}{2},1]})$, we get $R(f) = 16e$ as well as $R(\Pi_\lambda f) = R(3e) = (3^\kappa)\,\dot{}$, so that we arrive at the opposite inequality

$$[R(\Pi_\lambda f)](\lambda) = \tfrac{1}{2}81 + \tfrac{1}{2}9 = 45 > 16 = [\Pi_\lambda(R(f))](\lambda). \qquad \bowtie$$

Remark: From the proof of the preceding theorem we see that

$$[\Pi_P(L*F)]\hat{\ }(z) = E_{\hat{P}}(L \circ \hat{F}|\hat{\mathcal{A}})(z) \quad \text{holds} \quad \hat{P}\text{-almost surely.}$$

Furthermore, if $\hat{F} \in C^B(Z)$ is *Bochner* integrable with respect to $\hat{P} \in M(Z)$ (which holds true, e.g. if B^* is separable, cf. the proof of Theorem 1.14), then we infer

$$[^B(\Pi_P)F]\hat{\ }(z) = E_{\hat{P}}(\hat{F}|\hat{\mathcal{A}})(z) \quad \hat{P}\text{-almost surely,}$$

the expression on the right-hand side denoting the conditional expectation in *Bochner's* sense (see [*Scalora* 1961]). Indeed, this relation is implied by

$$\langle x, \int_U (E_{\hat{P}}(\hat{F}|\hat{\mathcal{A}}) - [^B(\Pi_P)F]\hat{\ })\,d\hat{P}\rangle = \int_U (\langle x, E_{\hat{P}}(\hat{F}|\hat{\mathcal{A}})\rangle - \langle x, {}^B(\Pi_P)F\rangle)\,d\hat{P}$$

$$= \int_U (E_{\hat{P}}(\widehat{\langle x, F\rangle}|\hat{\mathcal{A}}) - [\Pi_P(\langle x, F\rangle)]\hat{\ })\,d\hat{P} = 0$$

for all $U \in \hat{\mathcal{A}}$ and all $x \in B$. In this case, Theorem 2.7 follows from the classical version of *Jensen's* inequality for *Banach* space valued random elements

$$L(E_{\hat{P}}(\hat{F}|\hat{\mathcal{A}})(z)) \leq E_{\hat{P}}(L \circ \hat{F}|\hat{\mathcal{A}})(z) \quad \hat{P}\text{-almost surely}$$

(see [*To Ting On/ Yip Kai Wing* 1975], [*Kozek/Suchanecki* 1980] or [*Bomze* 1984]). However, since the equation

$$\int_U E_{\hat{P}}(\langle x, \hat{F}\rangle|\hat{\mathcal{A}})\,d\hat{P} = \int_U \langle x, {}^B(\Pi_P)F\rangle\,d\hat{P}$$

holds for all $U \in \hat{\mathcal{A}}$ and all $x \in B$, independently of whether or not $\hat{F}$ is *Bochner* integrable with respect to $\hat{P}$, we can in any case view at $[^B(\Pi_P)F]\hat{\ }$ as a conditional expectation, namely in context of the $\sigma(B^*, B)$-*Pettis* integral of $\hat{\mathcal{M}}$-measurable functions defined on Z taking values in B^* (see the remark preceding Proposition 1.8 and the proof of Theorem 1.14). $\triangle$

3. Sufficiency

In statistical theories of estimation and testing hypotheses, the notion of sufficiency plays an important role (see, e.g. [*Schmetterer* 1974]; [*Strasser* 1985]). It represents the abstraction of *Fisher*'s [1922, p.316; 1925, p.713] idea to condense the information relevant for an experiment in a statistic. Formally, this means that the conditional expectation of the observations, given the sufficient statistic, is independent of the underlying probability measure [*Neyman/Pearson* 1936]. Further generalization is obtained by transition from sub-σ-fields induced by a statistic to arbitrary sub-σ-fields $\mathcal{C}$ of $\mathcal{F}$ (cf. [*Bahadur/Lehmann* 1955]): $\mathcal{C}$ is called sufficient for $\mathcal{P}$, if for all $\varphi \in X$ there is a version of $E_P(\varphi|\mathcal{C})$ that is independent of $P \in \mathcal{P}$.

Mainly to clarify the relations between sufficiency and decision theory [*Wald* 1950], *LeCam* [1964; in particular pp.1419f.] extended the notion of sufficiency to the L- and M-spaces of an experiment (see also chapter 5 in [*LeCam* 1986]). In the sequel it will turn out that this further abstraction – besides its decision theoretic interpretation – is of central importance for the treatment of invariance and optimality in undominated experiments.

The considerations in the preceding chapter suggest to apply the notion of sufficiency to the experiment $\hat{\mathcal{E}} = (Z, \hat{\mathcal{M}}, \hat{\mathcal{P}})$ and to transfer it by this means to $\mathcal{L}$ or $\mathcal{M}$, respectively. In section 3.1., we shall show that this approach, on one hand, yields the same concepts as constructed by *LeCam*'s [1964] method and, on the other hand, is closely related to the generalized conditional expectations Π_P from Theorem 2.4. Moreover we specify in section 3.2. a criterion for sufficiency which can be viewed as generalization of the theorem due to *Halmos/Savage* [1949] valid for dominated experiments. This section is also devoted to further investigations of sufficiency properties, which enable us to prove, in section 3.3., some results especially important for chapter 5, e.g. the existence of a smallest sufficient sub-algebra – a further analogy to the dominated case in the classical setting. Finally, section 3.4 deals with the role sufficiency plays for *Banach* space valued generalized random elements.

3. Sufficiency

The following result shows that continuous versions of the conditional expectation, given a sub-σ-field $\hat{A}$ of the system $\hat{M}$ of *Baire* sets in Z, are already uniquely determined, provided they do not depend on the measures $\hat{P}$, $P \in \mathcal{P}$ (cf. Proposition 2.3).

Proposition 3.1:

Let A be a linear sub-space of M and $f \in M$. An element $h \in A$ fulfilling

$$(f \cdot g)(P) = (h \cdot g)(P) \quad \text{for all } g \in A \text{ and all } P \in \mathcal{P}$$

is already uniquely determined by this relation.

Proof: Suppose $\{h, h'\} \subseteq A$ fulfills $(h \cdot g)(P) = (f \cdot g)(P) = (h' \cdot g)(P)$ for all $g \in A$ and all $P \in \mathcal{P}$. Taking $g = h - h' \in A$ we thus have $[S*(h - h')](P) = (h \cdot g)(P) - (h' \cdot g)(P) = 0$ for all $P \in \mathcal{P}$. Proposition 1.5(c) and Proposition 1.3(b) now yield $h = h'$. $\qquad\square$

Now we give three equivalent definitions of sufficiency:

Theorem 3.2 and Definition:

Let A be a $\|.\|$-closed sub-algebra in M containing e. Then the following assertions are equivalent:

(a) $\hat{A}$ is sufficient for $\{\hat{P} : P \in \mathcal{P}\}$ in $(Z, \hat{M})$;

(b) for all $f \in M_+$, it holds that $\bigcap_{P \in \mathcal{P}} S_A(f, P) \neq \emptyset$;

(c) there is a linear, positive map $\Pi : M \to A$ fulfilling (a) and (b) in Theorem 2.4 for all $P \in \mathcal{P}$ (i.e. in the situation described by Theorem 2.4 one may choose $\Pi_P = \Pi$ for all $P \in \mathcal{P}$).

If one (and therefore all) of the statements in (a), (b), or (c), are true, we call A "sufficient". The map Π in (c) is uniquely determined by A and fulfills

$$\bigcap_{P \in \mathcal{P}} S_A(f, P) = \{\Pi f\} \quad \text{as well as} \quad \Pi f(P) = f(P) \quad \text{for all } P \in \mathcal{P},$$

if $f \in M_+$. This map Π is called "sufficient projection onto A".

Proof: (a) $\Rightarrow$ (b): let $f \in \mathcal{M}_+$ and $r : Z \to [0, \|f\|]$ be a version of $E_{\hat{P}}(\hat{f}|\hat{\mathcal{A}})$ that is independent of $\hat{P}$. Since r is $\hat{\mathcal{A}}$-measurable, Theorem 2.2 ensures the existence of an element $h \in \mathcal{A}_+$ fulfilling

$$\hat{h}(z) = r(z) = E_{\hat{P}}(\hat{f}|\hat{\mathcal{A}})(z) \quad \hat{P}\text{-almost surely.}$$

Now Proposition 2.3 yields $h \in \bigcap_{P \in \mathcal{P}} \mathcal{S}_{\mathcal{A}}(f, P)$.

(b) $\Rightarrow$ (c): again, let $f \in \mathcal{M}_+$. By Proposition 3.1 we infer that $\bigcap_{P \in \mathcal{P}} \mathcal{S}_{\mathcal{A}}(f, P)$ contains exactly one element which we denote by Πf. Defining $\Pi f = \Pi f_+ - \Pi f_-$ for all $f \in \mathcal{M}$, we obtain a linear, positive map $\Pi : \mathcal{M} \to \mathcal{A}$ which has the properties (a) and (b) from Theorem 2.4. This follows from Proposition 2.3 and from the relation $\Pi g = g$ for all $g \in \mathcal{A}$, which holds because of $g_\pm \in \bigcap_{P \in \mathcal{P}} \mathcal{S}_{\mathcal{A}}(g_\pm, P)$ for any $g \in \mathcal{A}$.

(c) $\Rightarrow$ (a): if $r : Z \to \mathbb{R}$ is bounded and $\hat{\mathcal{M}}$-measurable and if $U' \in \hat{\mathcal{A}}$, then choose $f \in \mathcal{M}$ and a set $U \in \mathcal{B}_{\mathcal{A}}$ such that

$$\hat{f}(z) = r(z) \ \hat{\mu}\text{-almost surely} \quad \text{as well as} \quad \hat{\mu}(U \triangle U') = 0$$

holds for all $\mu \in \mathcal{L}$. From the relation $s_U \in \mathcal{A}$ and from Proposition 2.5 we infer

$$\int_{U'} r \, d\hat{P} = \int_{U'} \hat{f} \, d\hat{P} = (s_U \cdot f)(P) = (s_U \cdot \Pi f)(P) = \int_{U'} \widehat{(\Pi f)} \, d\hat{P},$$

showing that $\widehat{(\Pi f)}$ is a version of $E_{\hat{P}}(r|\hat{\mathcal{A}})$ which is independent of $\hat{P}$. $\qquad\square$

Remark: Even if it exists, the sufficient projection Π need not, in general, coincide with the maps Π_P constructed in the proof of Theorem 2.4: if, for instance, $P(A) = 0$ holds for some $A \in \mathcal{F} \setminus \mathcal{N}$, then (for $\mathcal{A} = \mathcal{M}$) we obtain $\Pi_P e \le e - 1_A < e = \Pi e$. Furthermore, note that if $\mathcal{C}$ is a sufficient sub-σ-field of $\mathcal{F}$ and if $\mathcal{A}$ is constructed as in the remark following Proposition 2.3, then $\mathcal{A}$ is also sufficient: indeed, for $f \in \mathcal{M}_+$ there are, by Proposition 1.2, $\mathcal{F}$-measurable functions $\varphi_i : \Omega \to [0, \|f\|]$ such that $\dot{\varphi}_i \to f$ w.r.t. $\sigma(\mathcal{M}, \mathcal{L})$ as $i \in I$. Moreover from sufficiency of $\mathcal{C}$ it follows that there exist versions $\psi_i : \Omega \to [0, \|f\|]$ of $E_P(\varphi_i|\mathcal{C})$ that are independent of P; this means $\dot{\psi}_i \in \bigcap_{P \in \mathcal{P}} \mathcal{S}_{\mathcal{A}}(\dot{\varphi}_i, P)$ for all $i \in I$. Now any $\sigma(\mathcal{M}, \mathcal{L})$-accumulation point h of the bounded net $(\dot{\psi}_i)_{i \in I}$ in $\mathcal{M}$, satisfies $h \in \mathcal{A}_+$ and therefore also $h \in \bigcap_{P \in \mathcal{P}} \mathcal{S}_{\mathcal{A}}(f, P)$ because of Proposition 1.5(b). Hence condition (b) from Theorem 3.2 is fulfilled (due to *Alaoglu*'s theorem, at least one such h exists; the subsequent Theorem 3.3 entails even $\dot{\psi}_i \to h$ w.r.t. $\sigma(\mathcal{M}, \mathcal{L})$ as $i \in I$). $\qquad\triangle$

3.2. Sufficiency characterizations and related properties

The important characterization of sufficiency which we now specify resembles in some sense
the following result of *Halmos/Savage* [1949] for the case $\mathcal{P} \stackrel{\cdot}{\cong} \mathcal{Q}$: a sub-$\sigma$-field $\mathcal{C}$ of $\mathcal{F}$ is
sufficient if and only if there are $\mathcal{C}$-measurable versions of the *Radon/Nikodym*-densities
$\psi_P = \frac{dP}{dQ}$ (compare ψ_P with $\Lambda\mu$ below; see also [*Luschgy* et al. 1988]).

For convenient notation, we denote by

$$\mathcal{L}(\mathcal{A}) := \{\sum_{i=1}^{n} g_i \cdot P_i : g_i \in \mathcal{A}, P_i \in \mathcal{P}, 1 \le i \le n; n \in \mathbb{N}\},$$

whenever $\mathcal{A} \subseteq \mathcal{M}$.

Theorem 3.3:

    If $\mathcal{A}$ is a $\|.\|$ -closed sub-algebra containing e, then the following assertions are equivalent:

(a) $\mathcal{A}$ is sufficient;

(b) there exists a linear, positive map $\Pi : \mathcal{M} \to \mathcal{A}$ fulfilling
    $\Pi g = g$ for all $g \in \mathcal{A}$ and $(\Pi f)(P) = f(P)$ for all $f \in \mathcal{M}$ and all $P \in \mathcal{P}$;

(c) for all $\mu \in \mathcal{L}(\mathcal{A})$ it holds $\|\mu\| = \sup\{g(\mu) : g \in \mathcal{A}, \|g\| \le 1\}$;

(d) there exists a linear, positive map $\Lambda' : \mathcal{L}' \to \mathcal{L}(\mathcal{A})$ fulfilling $\Lambda' \circ \Lambda' = \Lambda'$ and
    $\Lambda' P = P$ for all $P \in \mathcal{P}$ as well as $\ker \Lambda' = \mathcal{A}^0$, which means

$$\Lambda'\mu = o \quad \text{if and only if} \quad f(\mu) = 0 \text{ for all } f \in \mathcal{A}.$$

If one (and hence all) of the statements in (a),(b),(c), and (d), are true, then the map
Π from (b) is the sufficient projection onto $\mathcal{A}$, and $\Pi = \Lambda^*$ holds, where $\Lambda : \mathcal{L} \to \mathcal{L}$
is the (unique) linear, positive map with $\Lambda|_{\mathcal{L}'} = \Lambda'$ from assertion (d). In particular
Π is $\sigma(\mathcal{M}, \mathcal{L})$-$\sigma(\mathcal{M}, \mathcal{L})$-continuous and $\mathcal{A} = \{f \in \mathcal{M} : \Pi f = f\}$ is $\sigma(\mathcal{M}, \mathcal{L})$-closed.
Moreover, this map Λ satisfies the following condition:
if $\mu \in \mathcal{L}'$ and if $\alpha_i \ge 0, P_i \in \mathcal{P}, 1 \le i \le n$, are chosen such that $|\mu| \le Q := \sum_{i=1}^{n} \alpha_i P_i$,
and if furthermore $g \in \mathcal{A}$ fulfills

$$\hat{g}(z) = \frac{d\hat{\mu}|_{\hat{\mathcal{A}}}}{d\hat{Q}|_{\hat{\mathcal{A}}}}(z) \quad \hat{Q}\text{-almost surely holds,}$$

then $\Lambda\mu = g \cdot Q$.

46

Proof: (a) $\Rightarrow$ (b) is an immediate consequence of Theorem 3.2.

(b) $\Rightarrow$ (c): let $\mu = \sum_{i=1}^{n} g_i \cdot P_i \in \mathcal{L}(\mathcal{A})$. Proposition 2.5 and Corollary 1.6 yield

$$f(\mu) = \sum_{i=1}^{n}(g_i \cdot f)(P_i) = \sum_{i=1}^{n}(g_i \cdot \Pi f)(P_i) = (\Pi f)(\mu),$$

which in turn entails (putting $g = \Pi f \in \mathcal{A}$)

$$\|\mu\| \le \sup\{g(\mu) : g \in \mathcal{A}, \|g\| \le 1\},$$

since $\|\Pi f\| \le \|f\|$ holds due to Proposition 2.5. The reverse inequality $\|\mu\| \ge \sup\{g(\mu) : g \in \mathcal{A}, \|g\| \le 1\}$ is trivial.

(c) $\Rightarrow$ (d): consider a measure $\mu \in (\mathcal{L}')_+$ with $\mu \le Q := \sum_{i=1}^{n}\alpha_i P_i$, where $\alpha_i \ge 0$ and $P_i \in \mathcal{P}$, $1 \le i \le n$. According to Theorem 2.2 there is an element $g \in \mathcal{A}_+$ such that

$$\hat{g}(z) = \frac{d\hat{\mu}|_{\hat{\mathcal{A}}}}{d\hat{Q}|_{\hat{\mathcal{A}}}}(z) \quad \hat{Q}\text{-almost surely holds.}$$

Thus $g \cdot Q \in \mathcal{L}(\mathcal{A})$ fulfills

$$f(g \cdot Q) = \int \hat{f} \cdot \hat{g}\, d\hat{Q} = \int \hat{f}\, d\hat{\mu} = f(\mu) \quad \text{for all } f \in \mathcal{A},$$

since $\hat{f}$ is $\hat{\mathcal{A}}$-measurable. However this property determines $g \cdot Q$ uniquely, because for $g' \in \mathcal{A}, Q' = \sum_{j=1}^{m}\beta_j P_j'$ we have $g \cdot Q - g' \cdot Q' \in \mathcal{L}(\mathcal{A})$ and $f(g \cdot Q - g' \cdot Q') = f(\mu) - f(\mu) = 0$ for all $f \in \mathcal{A}$, yielding, by assumption, $\|g \cdot Q - g' \cdot Q'\| = 0$. Hence the map

$$\begin{aligned} \Lambda'_+ : \quad \mu &\mapsto g \cdot Q \\ (\mathcal{L}')_+ &\to \mathcal{A}_+ \end{aligned}$$

is well defined, additive, and positively homogeneous on $(\mathcal{L}')_+$, thus we may extend it in an unique way to a linear, positive map Λ' defined on $\mathcal{L}'$. In the considerations above, we proved that $g \cdot Q$ is unique. By consequence, $\{\Lambda'\mu, P\} \subseteq \mathcal{L}(\mathcal{A})$ for all $\mu \in \mathcal{L}'$ and all $P \in \mathcal{P}$ implies

$$\Lambda'(\Lambda'\mu) = \Lambda'\mu \quad \text{as well as} \quad \Lambda' P = P.$$

Finally, the assumption together with $f(\Lambda'\mu) = f(\mu)$ and $\Lambda'\mu \in \mathcal{L}(\mathcal{A})$ for all $f \in \mathcal{A}$ and all $\mu \in \mathcal{L}'$ yield the relation $\ker \Lambda' = \mathcal{A}^0$.

(d) $\to$ (a): due to Proposition 1.1 the map $\Lambda' : \mathcal{L}' \to \mathcal{L}(\mathcal{A})$ has a unique linear, positive extension on $\mathcal{L}$, which we denote by Λ. For continuity reasons, this map fulfills $\Lambda(\mathcal{L}) \subseteq \mathcal{L}$, and also

$$\Lambda \circ \Lambda = \Lambda \quad \text{and} \quad g \circ \Lambda = g \text{ for all } g \in \mathcal{A},$$

the last equation being a consequence of $g(\Lambda\mu - \mu) = 0$ for all $\mu \in \mathcal{L}'$, which in turn follows by

$$\Lambda(\Lambda\mu - \mu) = o\,.$$

The adjoint map

$$\Pi = \Lambda^* : \mathcal{M} \to \mathcal{M}$$

clearly is linear and positive, too. For $f \in \mathcal{M}$ and $g \in \mathcal{A}$ we have

$$((\Pi f) \cdot g)(P) = (\Pi f)(g \cdot P) = (f \cdot g)(P) \quad \text{for all } P \in \mathcal{P}\,,$$

since $\Lambda(\Lambda(g \cdot P) - g \cdot P) = o$ entails $f(\Lambda(g \cdot P)) = f(g \cdot P)$. Furthermore we note that $\mathcal{A} \subseteq \{f \in \mathcal{M} : \Pi f = f\}$ holds. Now suppose there were an element $f = \Pi f \notin \mathcal{A}$. Since $\mathcal{A}$ is a $\sigma(\mathcal{M}, \mathcal{L})$-closed linear sub-space of $\mathcal{M}$, there were a measure $\mu \in \mathcal{L}$ such that

$$f(\Lambda\mu) = (\Pi f)(\mu) = f(\mu) > 0 = g(\mu) \quad \text{for all } g \in \mathcal{A}\,,$$

which in turn, by assumption, would imply $\Lambda\mu = o$ whence we could deduce the contradiction $f(\Lambda\mu) = 0$. Therefore $\mathcal{A} = \{f \in \mathcal{M} : f = \Pi f\}$ is true. Sufficiency of $\mathcal{A}$ now follows from Theorem 3.2(c). The remaining assertions are evident from the above arguments. $\square$

Remark: The relation $\Pi = \Lambda^*$ is essentially equivalent to the assertion of Lemma 2 in [*LeCam* 1964, p.1433], while the implication (b) $\Rightarrow$ (c) sharpens a result of Proposition 11 [*LeCam* 1964, p.1437], which investigates the relation between sufficiency and pairwise sufficiency (cf. [*Siebert* 1979] and chapter 5 in [*LeCam* 1986]). $\triangle$

The following Proposition 3.4 summarizes several sufficiency properties; its assertions were stated – and proved by somewhat different arguments – already in [*LeCam* 1964, pp.1435ff.].

Proposition 3.4:
(a) If $\mathcal{A}$ is sufficient and Π is the sufficient projection onto $\mathcal{A}$, then
    (i) $\Pi : \mathcal{M} \to \mathcal{M}$ is a linear, positive map fulfilling $\Pi \circ \Pi = \Pi$ and $\Pi e = e$;
    (ii) unbiasedness: $(\Pi f)(P) = f(P)$ for all $f \in \mathcal{M}$ and all $P \in \mathcal{P}$;
    (iii) *Reynold's* property: $\Pi(f \cdot g) = (\Pi f) \cdot g$ for all $f \in \mathcal{M}$ and all $g \in \mathcal{A}$;
        in particular we have $\Pi g = g$ for all $g \in \mathcal{A}$;
    (iv) Π is $\sigma(\mathcal{M}, \mathcal{L})$-$\sigma(\mathcal{M}, \mathcal{L})$-continuous;
    (v) *Jensen's* inequality: for a CB function L on a *Banach* space B and every $F \in \mathcal{M}^B$ we have $L * (\Pi^B F) \leq \Pi(L * F)$.

48

(b) Every map Π fulfilling (i) and (ii) in assertion (a) is already a sufficient projection, namely onto the $\sigma(\mathcal{M},\mathcal{L})$-closed sub-algebra $\mathcal{A} = \Pi(\mathcal{M})$; in particular $\mathcal{A}$ is uniquely determined by Π. Conversely, for any map $\Pi : \mathcal{M} \to \mathcal{A}$, $\mathcal{A}$ being a $\|.\|$-closed sub-vector-lattice of $\mathcal{M}$, property (i) is also a consequence of properties (ii) and (iii) in (a).

(c) Let $\Delta : \mathcal{M} \to \mathcal{M}$ be a linear, positive map that is $\sigma(\mathcal{M},\mathcal{L})$-$\sigma(\mathcal{M},\mathcal{L})$-continuous and that fulfills $\Delta e = e$ as well as $(\Delta f)(P) = f(P)$ for all $f \in \mathcal{M}$ and all $P \in \mathcal{P}$. For $(f,\mu) \in \mathcal{M} \times \mathcal{L}$ define

$$W_n(f,\mu) := \frac{1}{n} \sum_{k=0}^{n-1} (\Delta^k f)(\mu), \quad n \in \mathbb{N}.$$

Then $|W_n(f,\mu)| \leq \|f\|\|\mu\|$ holds for all $n \in \mathbb{N}$. Furthermore, every accumulation point W of the sequence $(W_n)_{n\in\mathbb{N}}$ with respect to the product topology in $\mathbb{R}^{\mathcal{M}\times\mathcal{L}}$ gives rise to a map $\Pi : f \mapsto W(f,.)$ from $\mathcal{M}$ into itself, sharing the properties (i) and (ii) in (a).

Proof: (a) is a consequence of Theorems 3.2 and 3.3 as well as of Propositions 2.5 and 2.6.

(b): because of $\Pi \circ \Pi = \Pi$ we have $\mathcal{A} = \{f \in \mathcal{M} : \Pi f = f\}$, hence $\mathcal{A}$ is a $\|.\|$-closed linear sub-space of $\mathcal{M}$ containing e. $\mathcal{A}$ is a sub-lattice, too: indeed, for $f \in \mathcal{A}$ positivity of Π yields $\Pi(f_+) \geq \Pi f = f$ and $\Pi(f_+) \geq o$, thus $\Pi(f_+) \geq f_+$ holds. Furthermore we have $[\Pi(f_+)](P) = f_+(P)$ for all $P \in \mathcal{P}$, which entails by Proposition 1.2(b) $\Pi(f_+) = f_+$, whence $f_+ \in \mathcal{A}$ follows.

Suppose now, conversely, that $\mathcal{A}$ is a $\|.\|$-closed sub-vector lattice of $\mathcal{M}$ and that an arbitrary map $\Pi : \mathcal{M} \to \mathcal{A}$ has the properties specified in (ii) and (iii) from (a). By Proposition 3.1, Π is linear, whence it follows that $\Pi g = g$ holds for all $g \in \mathcal{A}$ as well as $\Pi \circ \Pi = \Pi$. For $f \in \mathcal{M}_+$ Proposition 2.3 implies

$$\widehat{\Pi f}(z) = E_{\hat{P}}(\hat{f}|\hat{\mathcal{A}})(z) \geq 0 \quad \hat{P}\text{-almost surely.}$$

Thus for $h := (\Pi f)_+$ we on one hand have $h \in \mathcal{A}$, according to Proposition 2.1. On the other hand

$$\hat{h}(z) = \max\{\widehat{\Pi f}(z), 0\} = E_{\hat{P}}(\hat{f}|\hat{\mathcal{A}})(z) \quad \hat{P}\text{-almost surely},$$

entailing, again by Proposition 2.3, $(f \cdot g)(P) = (h \cdot g)(P)$ for all $P \in \mathcal{P}$, whence we obtain via Proposition 3.1 the relation $\Pi f = h \in \mathcal{M}_+$, which proves positivity of Π.

(c): from the inequalities $|\Delta f| \leq \Delta |f| \leq \Delta(\|f\|e) = \|f\|e$ we derive $\|\Delta f\| \leq \|f\|$, thus $|W_n(f, \mu)| \leq \|f\| \|\mu\|$ is true. Theorem 1.7 guarantees the existence of at least one map W having the claimed properties so that the assertion is not void. Suppose now that

$$W_{n_i}(f, \cdot) = \frac{1}{n_i} \sum_{k=0}^{n_i - 1} \Delta^k f \to \Pi f \quad \text{w.r.t. } \sigma(\mathcal{M}, \mathcal{L}) \text{ as } i \in I$$

is true for all $f \in \mathcal{M}$, where $(n_i)_{i \in I}$ is a suitably chosen sub-net of $\mathbb{N}$. By consequence, we conclude

$$(\Pi f)(P) = \lim_{i \in I} W_{n_i}(f, P) = f(P) \quad \text{for all } f \in \mathcal{M} \text{ and all } P \in \mathcal{P}.$$

Since Δ is $\sigma(\mathcal{M}, \mathcal{L})$-$\sigma(\mathcal{M}, \mathcal{L})$-continuous, we obtain for all $m \in \mathbb{N}$

$$\Delta^m \Pi f(\mu) = \lim_{i \in I} \frac{1}{n_i} \sum_{k=0}^{n_i - 1} \Delta^{k+m} f(\mu)$$

$$= \lim_{i \in I} \frac{1}{n_i} \sum_{k=0}^{n_i - 1} \Delta^k f(\mu) = (\Pi f)(\mu) \quad \text{for all } \mu \in \mathcal{L},$$

which shows $\Pi \circ \Pi = \Pi$. Finally we arrive via $W_n(f, \mu) = e(\mu)$ at the evident equality

$$(\Pi e)(\mu) = \lim_{i \in I} W_{n_i}(e, \mu) = e(\mu) \quad \text{holding for all } f \in \mathcal{M} \text{ and all } \mu \in \mathcal{L}. \qquad \square$$

3.3. Intersections; the minimal sufficient sub-algebra

The subsequent considerations show that sufficiency is a property preserved under intersection operations and, furthermore, enable us to calculate explicitly the sufficient projection onto the intersection of sufficient sub-algebras. To this end we next prove a relation which strongly resembles the assertion of Theorem 3 in [*Burkholder/Chow* 1961]:

Theorem 3.5:

Let $\mathcal{A}_i$ be sufficient and denote by Π_i the sufficient projections onto $\mathcal{A}_i$, $1 \leq i \leq m$. Put $\mathcal{A} = \bigcap_{i=1}^{m} \mathcal{A}_i$. Then

(a) $\mathcal{A}$ is sufficient; if Π denotes the sufficient projection onto $\mathcal{A}$, then

(b) for $m = 2$ we have

$$\frac{1}{n} \sum_{k=0}^{n-1} (\Pi_1 \circ \Pi_2)^k f \to \Pi f \text{ w.r.t. } \sigma(\mathcal{M}, \mathcal{L}) \text{ as } n \to \infty;$$

(c) for arbitrary m we have, provided that $\Pi_i \circ \Pi_j = \Pi_j \circ \Pi_i$ holds for all $\{i, j\} \subseteq \{1, \ldots, m\}$:

$$\Pi = \Pi_1 \circ \ldots \circ \Pi_m .$$

Proof: We begin by proving the key assertion (b). The map $\Delta := \Pi_1 \circ \Pi_2 : \mathcal{M} \to \mathcal{M}$ is linear, positive, $\sigma(\mathcal{M}, \mathcal{L})$-$\sigma(\mathcal{M}, \mathcal{L})$-continuous, and fulfills $\Delta e = e$ as well as $(\Delta f)(P) = f(P)$ for all $P \in \mathcal{P}$ and all $f \in \mathcal{M}$. According to Proposition 3.4 there is a sufficient sub-algebra $\mathcal{A}'$ with sufficient projection Π' and a sub-net $(n_i)_{i \in I}$ of $\mathbb{N}$ such that

$$\frac{1}{n_i} \sum_{k=0}^{n_i-1} \Delta^k f \to \Pi' f \text{ w.r.t. } \sigma(\mathcal{M}, \mathcal{L}) \text{ as } i \in I$$

for all $f \in \mathcal{M}$. Now $\Delta g = g$ for all $g \in \mathcal{A}'$ entails $\mathcal{A} \subseteq \{f \in \mathcal{M} : \Pi' f = f\} = \mathcal{A}'$. We now show that the converse inclusion $\mathcal{A}' \subseteq \mathcal{A}$ holds also: suppose that $f \in \mathcal{A}'$ and $g := \Pi_2 f \in \mathcal{A}_2$. Then

$$(g \cdot f)(P) = [(\Pi_2 f) \cdot f](P) = [(\Pi_2 f) \cdot (\Pi_2 f)](P) = (S * g)(P).$$

Furthermore, the equality $\Pi_1 y = \Delta f = \Delta \Pi' f = \Pi' f = f$ implies by Proposition 3.4(a)(v)

$$(S * f)(P) = (S * \Pi_1 g)(P) \leq [\Pi_1 (S * g)](P) = (S * g)(P),$$

whence we arrive at

$$[S*(f - g)](P) = (S*f)(P) + (S*g)(P) - 2(f \cdot g)(P) \leq 2[(S*g) - (f \cdot g)](P) = 0$$

for all $P \in \mathcal{P}$. From this result we infer via Propositions 1.3(b) and 1.5(c) the equality $f = g \in \mathcal{A}_2$. Evidently we also have $f = \Pi_1 g \in \mathcal{A}_1$, and thus $f \in \mathcal{A}$, which finally shows $\mathcal{A} = \mathcal{A}'$. By Proposition 3.4(b), $\mathcal{A}$ determines Π uniquely. Therefore we conclude $\Pi' = \Pi$. Hence there is exactly one accumulation point of the sequence $(\frac{1}{n} \sum_{k=0}^{n-1} \Delta^k)_{n \in \mathbb{N}}$ with respect to the product topology in $\mathbb{R}^{\mathcal{M} \times \mathcal{L}}$, namely Π. This proves assertion (b). Assertion (a) is derived from (b) by induction on m, whereas assertion (c) is implied by Proposition 3.4(b) with the help of the relation $\mathcal{A} = (\Pi_1 \circ \ldots \circ \Pi_m)(\mathcal{M})$. $\square$

Theorem 3.6:

Let $(\mathcal{A}_i)_{i \in I}$ be a (downwards) directed net of sufficient sub-algebras; this means that $\mathcal{A}_i \subseteq \mathcal{A}_j$ holds whenever the index j precedes the index i w.r.t. the partial ordering in I. Denote by Π_i the sufficient projection onto $\mathcal{A}_i$, $i \in I$. Then the intersection $\mathcal{A} = \bigcap_{i \in I} \mathcal{A}_i$ is sufficient and

$$\Pi_i f \to \Pi f \text{ w.r.t. } \sigma(\mathcal{M}, \mathcal{L}) \text{ as } i \in I,$$

if Π is the sufficient projection onto $\mathcal{A}$.

Proof: For $i \in I$ and $(f, \mu) \in \mathcal{M} \times \mathcal{L}$ define $W_i(f, \mu) := (\Pi_i f)(\mu)$. Because of $|W_i(f, \mu)| \leq \|f\|\|\mu\|$ and Theorem 1.7, there is a bilinear map W which is an accumulation point of the net $(W_i)_{i \in I}$ with respect to the product topology in $\mathbb{R}^{\mathcal{M} \times \mathcal{L}}$. Now define the linear functional Πf from $\mathcal{L}$ to $\mathbb{R}$ by $(\Pi f)(\mu) = W(f, \mu)$ and pick a suitable sub-net I' of I such that

$$\Pi_i f \to \Pi f \text{ w.r.t. } \sigma(\mathcal{M}, \mathcal{L}) \text{ as } i \in I'.$$

Since $\Pi_i f \in \mathcal{A}_i \subseteq \mathcal{A}_{i_0}$ holds whenever i_0 precedes i in I, and because $\mathcal{A}_{i_0}$ is $\sigma(\mathcal{M}, \mathcal{L})$-closed, we get $\Pi f \in \mathcal{A}$, due to cofinality of I' in I. Therefore $\mathcal{A}$ and Π satisfy the assumptions of Theorem 3.3(b), so that $\mathcal{A}$ is sufficient and Π is uniquely determined by $\mathcal{A}$. The remainder of the proof is completely analoguous to that of Theorem 3.5(b). $\square$

As a consequence of the preceding considerations, we obtain the fact, that arbitrary intersections of sufficient sub-algebras are again sufficient, and furthermore the conclusion that – in the sense of set inclusion – the smallest sufficient sub-algebra of $\mathcal{M}$ always exists (cf. [*LeCam* 1964, p.1436]):

Corollary 3.7:

(a) If $\mathcal{A}_i$ are sufficient sub-algebras for all $i \in I$, then $\bigcap_{i \in I} \mathcal{A}_i$ is sufficient, too;

(b) there exists the smallest sufficient sub-algebra $\mathcal{A}_m$ in $\mathcal{M}$.

Proof: (a) follows from Theorems 3.5 and 3.6.
(b): according to (a), the sub-algebra

$$\mathcal{A}_m := \bigcap_{\substack{\mathcal{A} \subseteq \mathcal{M} \\ \mathcal{A} \text{ sufficient}}} \mathcal{A}$$

is itself sufficient. $\qquad\square$

The intersection $\mathcal{C}_m$ of all sub-σ-fields $\mathcal{C}$ of $\mathcal{F}$, which are sufficient for $\mathcal{P}$ and contain $\mathcal{N}$, is called "minimal σ-field" of $\mathcal{E}$. For coherent, in particular for dominated experiments, $\mathcal{C}_m$ is sufficient for $\mathcal{P}$, a fact that is not true in the general case (see [Hasegawa/Perlman 1974], also [Halmos/Savage 1949] and [Siebert 1979]; cf. appendix):

Corollary 3.8:

If $\mathcal{E}$ is coherent, i.e. if $\dot{X} = \mathcal{M}$, then the minimal sub-σ-field $\mathcal{C}_m$ is sufficient for $\mathcal{P}$.

Proof: Let $\mathcal{C}$ be a sub-σ-field of $\mathcal{F}$ containing $\mathcal{N}$. Since

$$|\psi(\omega)| \leq \|\dot{\psi}\| \quad P\text{-almost surely holds for all } \psi \in X \text{ and all } P \in \mathcal{P},$$

and because $\|\dot{\varphi}_n - \dot{\varphi}\| \to 0$ as $n \to \infty$, the fact that all φ_n are $\mathcal{C}$-measurable yields $\mathcal{C}$-measurability of φ. Thus $\mathcal{A} := \dot{X}_\mathcal{C}$ is a $\|.\|$-closed sub-algebra of $\mathcal{M}$ (cf. remark following Proposition 2.3). $\mathcal{A}$ is sufficient if $\mathcal{C}$ is sufficient for $\mathcal{P}$: indeed, a version ψ of $E_P(\varphi|\mathcal{C})$ that is independent of P, $P \in \mathcal{P}$, fulfills

$$\dot{\psi} \in \bigcap_{P \in \mathcal{P}} \mathcal{S}_\mathcal{A}(f, P) .$$

$\mathcal{A}$ is hence even $\sigma(\mathcal{M}, \mathcal{L})$-closed (cf. Theorems 3.2 and 3.3), and thus contains $\mathcal{A}_m$. Therefore, by defining

$$\mathcal{C}'_m := \{A \in \mathcal{F} : 1_A \in \mathcal{A}_m\} ,$$

we get a sub-σ-field of $\mathcal{F}$ containing $\mathcal{N}$ (recall that $1_N = o \in \mathcal{A}_m$ holds for all $N \in \mathcal{N}$), which, due to the considerations specified at the begin of this proof, satisfies $\mathcal{C}'_m \subseteq \mathcal{C}_m$.

According to Proposition 2.1, a function $\varphi \in X$ is $\mathcal{C}_m$-measurable if and only if $\dot{\varphi} \in \mathcal{A}_m$. Therefore $\mathcal{C}'_m$ is sufficient for $\mathcal{P}$: indeed, since

$$\int_A \psi \, dP = (\Pi\dot{\varphi} \cdot \dot{1}_A)(P) = (\dot{\varphi} \cdot \dot{1}_A)(P) = \int_A \varphi \, dP \text{ for all } A \in \mathcal{C}'_m$$

holds if $\dot{\psi} = \Pi\dot{\varphi} \in \mathcal{A}_m$, ψ is a version of $E_P(\varphi|\mathcal{C})$ for all $P \in \mathcal{P}$ which does not depend on P. By consequence we finally have $\mathcal{C}'_m = \mathcal{C}_m$. $\qquad\square$

3.4. Banach space results

At the end of this chapter we specify – for the sake of completeness – a sufficiency result concerning generalized random elements in *Banach* spaces B of arbitrary dimension:

Theorem 3.9:

If $\mathcal{A}$ is a $\|.\|$-closed sub-algebra of $\mathcal{M}$ containing e, then the following assertions are equivalent:

(a) $\mathcal{A}$ is sufficient;

(b) there is a continuous linear map $\overline{\Pi} : \mathcal{M}^B \to \mathcal{A}^B$ fulfilling
   (i) $\overline{\Pi}F(P) = F(P)$ for all $F \in \mathcal{M}^B$ and all $P \in \mathcal{P}$,
   (ii) $\langle \overline{\Pi}F, y \rangle \geq o$, if $\langle F, y \rangle \geq o$ for all $F \in \mathcal{M}^B$ and all $y \in B^*$;
   (iii) $\overline{\Pi}F = F$ for all $F \in \mathcal{A}^B$;

(c) for every $\hat{F} \in C^B(Z)$ which is *Bochner* integrable with respect to all $\hat{P}$, $P \in \mathcal{P}$, there is a version of $E_{\hat{P}}(\hat{F}|\hat{\mathcal{A}})$ that does not depend on P.

Proof: (a) $\Rightarrow$ (b): Theorem 3.3 and Proposition 1.9(b) yield the existence of a linear map $\overline{\Pi} := \Pi^B : \mathcal{M}^B \to \mathcal{A}^B$ fulfilling $\langle \overline{\Pi}F, y \rangle = \Pi(\langle F, y \rangle)$ for all $y \in B^*$, where Π is the sufficient projection onto $\mathcal{A}$. Now the assertion claimed follows from the observations that

$$\langle \overline{\Pi}(g \cdot F), y \rangle = \Pi(\langle g \cdot F, y \rangle) = \Pi(g \cdot \langle F, y \rangle) = g \cdot \Pi(\langle F, y \rangle) = \langle g \cdot \overline{\Pi}F, y \rangle$$

and that

$$\langle (\overline{\Pi}F)(P), y \rangle = [\Pi(\langle F, y \rangle)](P) = \langle F, y \rangle(P)$$

holds for all $y \in B^*$.

(b) $\Rightarrow$ (a): choose $(x, y) \in B \times B^*$ such that $\langle x, y \rangle = 1$. Consider the functional

$$\Pi f : \mu \mapsto \langle \overline{\Pi}(f \cdot x), y \rangle(\mu) \, .$$
$$\mathcal{M} \to \mathbb{R}$$

Then $\Pi f \in \mathcal{A}$ and the map $f \mapsto \Pi f$ is linear and positive, since $f = \langle x, y \rangle f = \langle f \cdot x, y \rangle$ entails $(\Pi f)(\mu) = \langle \overline{\Pi}(f \cdot x), y \rangle(\mu) \geq 0$, if $f \in \mathcal{M}_+$ and $\mu \in \mathcal{L}_+$ holds. For $g \in \mathcal{A}$ we have $F := g \cdot x \in \mathcal{A}^B$, thus

$$(\Pi g)(\mu) = \langle \overline{\Pi} F, y \rangle(\mu) = \langle F, y \rangle(\mu) = g(\mu) \quad \text{for all } \mu \in \mathcal{L} \, ,$$

which means $\Pi g = g$. Now the assertion follows from Theorem 3.3.

(a) $\Rightarrow$ (c) is evident by the remark following Theorem 2.7.

(c) $\Rightarrow$ (a): let, again, $(x, y) \in B \times B^*$ be such that $\langle x, y \rangle = 1$ holds and choose a version R of $E_{\hat{P}}(\widehat{f \cdot y} | \hat{A})$ that does not depend on P (observe that $\widehat{f \cdot y} = \hat{f} \cdot y \in C^B(Z)$ is *Bochner integrable* with respect to all $\hat{P}$, $P \in \mathcal{P}$). Then we have

$$\langle x, R(z) \rangle = E_{\hat{P}}(\langle x, \hat{f} \cdot y \rangle | \hat{A})(z) = E_{\hat{P}}(\hat{f} | \hat{A})(z) \quad \hat{P}\text{-almost surely.}$$

Sufficiency of $\mathcal{A}$ now follows from the conclusions used in the proof of (a) $\Rightarrow$ (b) in Theorem 3.2. $\qquad \square$

Remark: If one – and thus all – of the conditions (a),(b),(c) in Theorem 3.9 are fulfilled, then $\hat{A}$ is sufficient for $\{\hat{P} : P \in \mathcal{P}\}$ in the sharpened sense that for all $\hat{F} \in C^B(Z)$ there is a version of the $\sigma(B^*, B)$-*Pettis* integral on $\hat{A}$ with respect to all $\hat{P}$, $P \in \mathcal{P}$, that does not depend on P, namely $\overline{\Pi} F$ (cf. the remark at the end of chapter 2). Evidently this property of $\mathcal{A}$ is equivalent to sufficiency. $\qquad \triangle$

4. Optimality and admissibility in unbiased estimation

We begin this chapter by recalling in section 4.1. some well-known facts about unbiased estimation in the classical setup. The remainder deals with the transfer of these results to generalized unbiased estimators in (undominated) experiments. For the sake of transparency, we proceed as follows: at first, we treat – except for the definitions of optimality and admissibility – the case $B = \mathbb{R}$ in sections 4.2. to 4.5.; in section 4.6., we shall reduce the case of a general *Banach* space B to the scalar case. We establish some auxiliary results in section 4.2., and then introduce and investigate in section 4.3. optimality and admissibility of generalized unbiased estimators. Section 4.4. is devoted to the relation between sufficiency, optimality, and admissibility. In section 4.5. we specify criteria for the existence of non-trivial generalized optimal estimators.

4.1. Unbiased estimation in the "classical" context: a summary

Let B be a *Banach* space and B^* be its topological dual. A "loss function" $\underline{L}$ on B (B^*), is a family $\underline{L} = (L_P)_{P \in \mathcal{P}}$ of CB (CB^*) functions L_P on B (B^*). $\underline{L}$ is called "strictly convex", if L_P are strictly convex for all $P \in \mathcal{P}$. For example,

$$\underline{S} := (S_P)_{P \in \mathcal{P}} \ , \ \text{where} \ S_P := S \quad \text{for all} \ P \in \mathcal{P},$$

is a strictly convex loss function on $B = \mathbb{R}$.

A parameter $\vartheta : \mathcal{P} \to B$ is called "unbiased estimable", if there is a function $\Phi \in X^B$ fulfilling $\vartheta(P) = \dot{\Phi}(P)$ for all $P \in \mathcal{P}$. In this case, Φ is an "unbiased estimator" for ϑ.

$$V^B := \{\Gamma \in X^B : \int \Gamma \, dP = o \ \text{for all} \ P \in \mathcal{P}\}$$

is the set of all unbiased estimators of zero $o \in B$. If $B = \mathbb{R}$, we write for short V instead of $V^{\mathbb{R}}$. Let Φ be an unbiased estimator for ϑ. The function $\Psi \in X^B$ is another unbiased estimator for ϑ if and only if $\Psi - \Phi \in V^B$ holds.

Blackwell [1947] noted (for $\underline{L} = \underline{S}$ and $B = \mathbb{R}$), that the risk is decreased by an application of the conditional expectation, given a sufficient sub-σ-field $\mathcal{C}$. Indeed, for any version Ψ of $E_P(\Phi|\mathcal{C})$, which does not depend on $P \in \mathcal{P}$, we on one hand have $\Psi - \Phi \in V^B$ and on the other hand obtain from *Jensen*'s inequality

$$\int L_P \circ \Psi \, dP \leq \int L_P \circ \Phi \, dP \quad \text{for all } P \in \mathcal{P}.$$

Of course it is desirable to obtain unbiased estimators with minimal risk compared to all unbiased alternatives. Therefore a function $\Phi \in X^B$ is called "$\underline{L}$-optimal", if

$$\int L_P \circ \Phi \, dP \leq \int L_P \circ (\Phi + \Gamma) \, dP \quad \text{for all } P \in \mathcal{P} \text{ and all } \Gamma \in V^B.$$

If an estimator Φ is $\underline{L}$-optimal for all loss functions $\underline{L}$, it is called "optimal".

An objective of the theory of unbiased estimation is to characterize optimality of estimators by means of measurability conditions. If we denote by

$$\mathcal{C}_0 := \{A \in \mathcal{F} : \int_A \gamma \, dP = 0 \text{ for all } P \in \mathcal{P} \text{ and all } \gamma \in V\}$$

$$= \{A \in \mathcal{F} : \int_A \Gamma \, dP = o \text{ for all } P \in \mathcal{P} \text{ and all } \Gamma \in V^B\},$$

then $\mathcal{C}_0$ is a sub-σ-field of $\mathcal{F}$ having the property that $\Phi \in X^B$ is measurable with respect to $\mathcal{C}_0$ if and only if Φ is optimal (see, e.g. [*Kozek* 1980]). In the case $B = \mathbb{R}$ *Bahadur* [1957] proved that ($\underline{S}$-)optimality yields $\mathcal{C}_0$-measurability; *Padmanabhan* [1970] and *Strasser* [1972] showed that, conversely, every $\mathcal{C}_0$-measurable $\varphi \in X$ already is optimal. Therefore $\mathcal{C}_0$ is called "optimal σ-field".

For the sake of lucidity, we in the following considerations confine ourselves to the case $B = \mathbb{R}$. Since $\text{Cov}_P(\varphi, \gamma) = 0$ holds true for all $\mathcal{C}_0$-measurable $\varphi \in X$ and all $(\gamma, P) \in V \times \mathcal{P}$, the equivalence of optimality and $\mathcal{C}_0$-measurability can be viewed as an obvious generalization of the "covariance method" due to *Rao* [1952], which states that $\varphi \in X$ is $\underline{S}$-optimal if and only if

$$\text{Cov}_P(\varphi, \gamma) = 0 \quad \text{holds for all } P \in \mathcal{P} \text{ and all } \gamma \in V.$$

An unbiased estimator φ of ϑ is $\underline{S}$-optimal if and only if φ is UMVUE, that is an unbiased estimator with uniformly minimal variance, of ϑ. This follows immediately from the equation $\mathrm{Var}_P \varphi = \int S \circ \varphi \, dP - [\vartheta(P)]^2$.

We call a sub-σ-field $\mathcal{C}$ of $\mathcal{F}$ "boundedly complete with respect to $\mathcal{P}$", if $\mathcal{N} \subseteq \mathcal{C}$ and $\dot{\gamma} = o$ holds for every $\mathcal{C}$-measurable $\gamma \in V$. Evidently $\mathcal{C}_0$ is boundedly complete. If $\mathcal{C}$ is a sub-σ-field of $\mathcal{F}$ that is sufficient for $\mathcal{P}$ and fulfills $\mathcal{N} \subseteq \mathcal{C}$, then $\mathcal{C}_0 \subseteq \mathcal{C}$ holds (see, e.g. [*Strasser* 1985, p.169]). Therefore any sub σ-field $\mathcal{C}$ of $\mathcal{F}$ that is boundedly complete with respect to and sufficient for $\mathcal{P}$, already coincides with $\mathcal{C}_0$; indeed, the reverse inclusion $\mathcal{C} \subseteq \mathcal{C}_0$, is derived as follows: for any $\gamma \in V$, a version $\psi \in X$ of $E_P(\gamma|\mathcal{C})$ that is independent of $P \in \mathcal{P}$, fulfills $\dot{\psi} = o$ due to the completeness of $\mathcal{C}$. This equality in turn entails for all $C \in \mathcal{C}$

$$\int_C \gamma \, dP \;=\; \int_C E_P(\gamma|\mathcal{C}) \, dP \;=\; \int_C \psi \, dP = 0 \ \text{ for all } P \in \mathcal{P}.$$

By consequence, we finally have $C \in \mathcal{C}_0$.

From these facts it is easy to conclude the following statement: the sufficiency of $\mathcal{C}_0$ is equivalent to the existence of a sub-σ-field of $\mathcal{F}$ which is complete with respect to and sufficient for $\mathcal{P}$ (recall that, by definition of $\mathcal{C}_0$, we have $\mathcal{N} \subseteq \mathcal{C}_0$). In that special situation – $\mathcal{C}_0$ sufficient – the equivalence of optimality and $\mathcal{C}_0$-measurability is already implied by the results due to *Lehmann/Scheffé* [1950] dealing with the equivalence of $\underline{S}$-optimality and $\mathcal{C}_0$-measurability (see, e.g. [*Strasser* 1985, p.167]).

A function $\varphi \in X$ is called "admissible for $\underline{L}$ (among unbiased alternatives)", if for any $\gamma \in V$ the relation

$$\int L_P \circ (\varphi + \gamma) \, dP \;\leq\; \int L_P \circ \varphi \, dP \quad \text{for all } P \in \mathcal{P}$$

yields already $\dot{\gamma} = o$. Any two functions $\varphi \in X$ and $\psi \in X$ which satisfy $\varphi - \psi \in V$, and which are admissible for a strictly convex loss function, are equivalent in the sense that $\dot{\varphi} = \dot{\psi}$ holds (see the remark following Theorem 4.14 below). Hence every optimal $\varphi \in X$ is admissible for any strictly convex loss function. Conversely, every estimator that is admissible for a strictly convex loss function is $\mathcal{C}_m$-measurable: indeed, the above considerations together with *Jensen*'s inequality imply $\dot{\varphi} = \dot{\psi}$, if ψ denotes a version of $E_P(\varphi|\mathcal{C})$, which does not depend on $P \in \mathcal{P}$, and if $\mathcal{C}$ is sufficient for $\mathcal{P}$ with $\mathcal{N} \subseteq \mathcal{C}$.

As a consequence of the last paragraph, we have $C_0 \subseteq C_m$ (cf. [*Kozek* 1980]). In general – even for dominated experiments $\mathcal{E}$ – the following inclusions are strict (see [*Bomze* 1986]):

$$\mathcal{N} \cup \{\Omega \setminus N : N \in \mathcal{N}\} \subset C_0 \subset C_m \, .$$

On the basis of the previous conclusions the equivalence of the following assertions is easily derived:

(a) C_0 is sufficient for $\mathcal{P}$;

(b) $C_0 = C_m$ and C_m is sufficient for $\mathcal{P}$;

(c) C_m is boundedly complete with respect to and sufficient for $\mathcal{P}$;

(d) there exists a sub-σ-field C of $\mathcal{F}$ which is boundedly complete with respect to and sufficient for $\mathcal{P}$.

If $\mathcal{E}$ is coherent – which by Corollary 3.8 implies that C_m is sufficient for $\mathcal{P}$ – then the assertions in (b) and (c) above can be sharpened in an obvious way. By contrast, for general experiments neither the relation $C_0 = C_m$ implies sufficiency of C_m (for example see [*Kozek* 1980] and [*Pitcher* 1965]); nor, conversely, sufficiency of C_m entails the equality $C_0 = C_m$.

To sum up, we can derive the equivalence of the following statements about $\varphi \in X$ provided that one – and thus all – of the assertions (a),(b),(c),(d) are true:

(i) φ is optimal;

(ii) φ is $\underline{L}$-optimal for all strictly convex loss functions $\underline{L}$;

(iii) φ is $\underline{S}$-optimal;

(iv) φ is admissible for all strictly convex loss functions $\underline{L}$;

(v) φ is admissible for $\underline{S}$;

(vi) φ is C_m-measurable;

(vii) φ is C_0-measurable.

If for arbitrary $\varphi \in X$ the conditions (i) and (ii) are equivalent, every strictly convex loss function $\underline{L}$, in particular $\underline{S}$, is "universal" in the sense that $\underline{L}$-optimality already implies optimality. In general there are strictly convex loss functions which are not universal, see [*Bednarek-Kozek/Kozek* 1978] and [*Bomze* 1986]. In the context of bounded estimators considered here, *Bahadur* [1957] proved that $\underline{S}$ is universal even if $C_0 \neq C_m$ holds. Among others, *Klebanov* [1972], *Schmetterer/Strasser* [1974], and *Kozek* [1980], investigated universal loss functions in the setup of possibly unbounded estimators.

The fact that (a) implies the equivalence of (iii) and (iv) above means that for every unbiased estimable parameter ϑ there is a UMVUE provided that C_0 is sufficient for $\mathcal{P}$. For dominated experiments $\mathcal{E}$ Bahadur [1957] showed that the converse of this statement is true.

The question whether or not there are non-trivial optimal estimators has been investigated for different classes of probability distributions, e.g. by *Bondesson* [1975], *Hoeffding* [1977a,b], and *Eberl* [1984]. For general experiments, *Bomze* [1986] in the dominated case $\mathcal{P} \cong Q \in ca(\Omega, \mathcal{F})_+$ proposed a criterion for the existence of non-trivial optimal estimators, which for finite sample spaces Ω was formulated and proved already by *Sapozhnikov* [1978]: the following assertions are equivalent

(α) $C_0 = \mathcal{N} \cup \{\Omega \setminus N : N \in \mathcal{N}\} = Q^{-1}(\{0, 1\})$;

(β) every optimal estimator is constant P-almost surely for all $P \in \mathcal{P}$;

(γ) $\mathcal{V} := \dot{V}$ is $\mathcal{P}$-irreducible.

Here, $\mathcal{P}$-irreducibility of $\mathcal{V}$ means that for any two closed linear sub-spaces $\mathcal{V}_1$ and $\mathcal{V}_2$ of $\mathcal{V}$ in $\mathcal{M} = L^\infty(Q)$ fulfilling

(1) $\mathcal{V} = \mathcal{V}_1 \oplus \mathcal{V}_2$ and

(2) $\dot{\gamma}_1 \cdot \dot{\gamma}_2 = o$ for all $(\gamma_1, \gamma_2) \in \mathcal{V}_1 \times \mathcal{V}_2$,

we have

$$\{\mathcal{V}_1, \mathcal{V}_2\} = \{\{o\}, \mathcal{V}\}.$$

Note that condition (2) above is equivalent to $Q([\mathcal{V}_1]_Q \cap [\mathcal{V}_2]_Q) = 0$, the symbol $[\mathcal{H}]_Q$ denoting the measurable Q-support of a sub-set $\mathcal{H} \subseteq L^1(Q)$, as defined in [*Bomze* 1986].

4.2. Auxiliary results

To begin with, we investigate the structure of the space of of all unbiased generalized estimators of zero 0, which we denote by

$$\mathcal{V} := \{v \in \mathcal{M} : v(P) = 0 \text{ for all } P \in \mathcal{P}\}.$$

For later use in section 4.6. let us note here, that for the spaces of all unbiased generalized B- $(B^*$-)valued estimators of zero $o \in B(B^*)$, we have the following characterization:

$$\{H \in \mathcal{M}^B : H(P) = o \text{ for all } P \in \mathcal{P}\} = \mathcal{V}^B \text{ and}$$
$$\{H \in {}^B\mathcal{M} : H(P) = o \text{ for all } P \in \mathcal{P}\} = {}^B\mathcal{V}.$$

Evidently we have $\dot{V}^B := \{\dot{\Gamma} : \Gamma \in V^B\} \subseteq (\dot{V})^B \subseteq \mathcal{V}^B \subseteq {}^{B^*}\mathcal{V}$, where we put $\dot{V} := \dot{V}^{\mathbb{R}} \subseteq \mathcal{M}$. The following observation clarifies in which sense $\mathcal{V}^B$ and ${}^{B^*}\mathcal{V}$ can be viewed as completion of $\dot{V}^B$. It deals with the relation between locally unbiased estimators and their generalization: for $\mathcal{Q} \in \Xi(\mathcal{P})$ we denote

$$V^B(\mathcal{Q}) := \{\Phi \in X^B : \int \Phi\, dP = o \text{ for all } P \in \mathcal{Q}\},\ \dot{V}^B(\mathcal{Q}) := \{\dot{\Phi} : \Phi \in V^B(\mathcal{Q})\} \text{ and}$$
$$\mathcal{V}(\mathcal{Q}) := \{f \in \mathcal{M} : f(P) = 0 \text{ for all } P \in \mathcal{Q}\}.$$

Then on one hand we have the identities

$$V^B = \bigcap_{\mathcal{Q} \in \Xi(\mathcal{P})} V^B(\mathcal{Q}),\ \dot{V}^B = \bigcap_{\mathcal{Q} \in \Xi(\mathcal{P})} \dot{V}^B(\mathcal{Q}),\ \text{and}\ \mathcal{V}^B = \bigcap_{\mathcal{Q} \in \Xi(\mathcal{P})} [\mathcal{V}(\mathcal{Q})]^B,$$

as well as, on the other hand, the following result:

Proposition 4.1:
   Let $\mathcal{Q} = \{P_i : 1 \leq i \leq n\} \in \Xi(\mathcal{P})$. Then $\dot{V}^B(\mathcal{Q})$ is

$$\sigma(\mathcal{M}, \mathcal{L})/\sigma(B, B^*)\text{-dense in } [\mathcal{V}(\mathcal{Q})]^B \text{ and also}$$
$$\sigma(\mathcal{M}, \mathcal{L})/\sigma(B^{**}, B^*)\text{-dense in } {}^{B^*}[\mathcal{V}(\mathcal{Q})].$$

Proof: It suffices to prove the claim for the case $B = \mathbb{R}$, i.e. for $V(\mathcal{Q}) := V^{\mathbb{R}}(\mathcal{Q})$. Proceeding indirectly, assume the contrary being true: in this case there were $f \in \mathcal{V}(\mathcal{Q})$, $\varepsilon > 0$ and $\mu \in \mathcal{L}$ such that

$$|\int \varphi \, d\mu - f(\mu)| \geq \varepsilon \quad \text{for all } \varphi \in V(\mathcal{Q}). \tag{$*$}$$

Hence we got $V(\mathcal{Q}) = V_1 \cup V_2$, where we have put

$$V_1 := \{\varphi \in V(\mathcal{Q}) : \dot{\varphi}(\mu) \leq f(\mu) - \varepsilon\} \text{ and}$$
$$V_2 := \{\varphi \in V(\mathcal{Q}) : \dot{\varphi}(\mu) \geq f(\mu) + \varepsilon\}.$$

Now if $V_1 = \emptyset$, then we had $\dot{\varphi}(\mu) = 0$ for all $\varphi \in V(\mathcal{Q})$, since $\alpha V(\mathcal{Q}) \subseteq V(\mathcal{Q})$ yielded $\alpha \dot{\varphi}(\mu) \geq f(\mu) + \varepsilon$ for all $\alpha \in \mathbb{R}$. Furthermore $(*)$ entailed

$$\mu \notin \mathcal{L}_{\mathcal{Q}} := \{\sum_{i=1}^{n} \alpha_i P_i : \alpha_i \in \mathbb{R}, \, 1 \leq i \leq n\}.$$

Because $\mathcal{L}_{\mathcal{Q}}$ is of finite dimension and hence a $\sigma(\mathcal{L}, \dot{X})$-closed linear sub-space of $\mathcal{L}$, there were a function $\varphi \in X$ fulfilling

$$\dot{\varphi}(\mu) > 0 = \dot{\varphi}(\nu) \quad \text{for all } \nu \in \mathcal{L}_{\mathcal{Q}}$$

[*Dunford/Schwartz* 1964, p.421]. But via $\dot{\varphi}(P) = 0$ for all $P \in \mathcal{Q}$ we concluded $\varphi \in V(\mathcal{Q})$, which in turn implied the contradiction

$$\dot{\varphi}(\mu) > 0 = \dot{\varphi}(\mu).$$

Thus we get $V_1 \neq \emptyset$; analogously one shows $V_2 \neq \emptyset$. On the other hand for $(\varphi, \psi) \in V_1 \times V_2$ we know

$$\alpha \varphi + (1 - \alpha)\psi \in V(\mathcal{Q}) \quad \text{for all } \alpha \in [0, 1],$$

entailing finally for

$$\alpha := \frac{\dot{\psi}(\mu) - f(\mu)}{\dot{\psi}(\mu) - \dot{\varphi}(\mu)}$$

the contradiction $[\alpha \varphi + (1 - \alpha)\psi](\mu) = f(\mu)$. $\qquad\square$

Next we specify a useful result, a proof of which would be trivial via the isomorphism $\mathcal{M} \simeq C(Z)$. For the sake of completeness we however give a short intrinsic argument:

62

Lemma 4.2:

Let $(g, \mu) \in \mathcal{M} \times \mathcal{L}_+$. Then the following assertions are equivalent:

(a) $(S*g)(\mu) = 0$;

(b) $|g|(\mu) = 0$;

(c) $(L*(f + g))(\mu) = (L*f)(\mu)$ for every convex $L : \mathbb{R} \to \mathbb{R}$ and every $f \in \mathcal{M}$.

Proof: (a) $\Rightarrow$ (b): for $\nu \in \mathcal{L}$ fulfilling $o < \nu \leq \mu$ Proposition 1.4(c) yields

$$[g(\nu)]^2 \leq \nu(\Omega)(S*g)(\nu) \leq \nu(\Omega)(S*g)(\mu) = 0 \, ,$$

which entails
$$|g|(\mu) = \sup\{g(\nu) - g(\mu - \nu) : o \leq \nu \leq \mu\} = 0 \, .$$

(b) $\Rightarrow$ (c): if $\mu_i \in \mathcal{L}_+, 1 \leq i \leq n$, fulfill $\sum_{i=1}^n \mu_i = \mu$ then

$$|g(\mu_i)| \leq |g|(\mu_i) \leq |g|(\mu) = 0 \, ,$$

yielding
$$\sum_{i=1}^n [\alpha_i(f + g)(\mu_i) + \beta_i \mu_i(\Omega)] = \sum_{i=1}^n [\alpha_i f(\mu_i) + \beta_i \mu_i(\Omega)]$$

and thus, via Proposition 1.4(d), the claimed assertion.

(c) $\Rightarrow$ (a): put $L = S$, $f = g$; then Proposition 1.5(c) implies $(S*g)(\mu) = 4(S*g)(\mu)$ and hence (a). $\qquad \square$

4.3. Optimality and admissibility

In this section we introduce the notions of optimality and admissiblity for generalized unbiased estimators, and give a criterion for optimality.

Definition:

(a) Let $\underline{L} = (L_P)_{P \in \mathcal{P}}$ be a loss function on B An element $F \in \mathcal{M}^B$ is called "$\underline{L}$-optimal", if for any $H \in \mathcal{V}^B$

$$(L_P * F)(P) \leq [L_P * (F + H)](P) \quad \text{for all } P \in \mathcal{P}.$$

Similarly, if $\underline{L}$ is a loss function on B^*, an element $F \in {}^B\mathcal{M}$ is called "$\underline{L}$-optimal", if the above relation is true for all $H \in {}^B\mathcal{V}$.

(b) An element $F \in \mathcal{M}^B \, ({}^B\mathcal{M})$ is called "optimal", if F is $\underline{L}$-optimal for all loss functions $\underline{L}$ on $B \, (B^*)$.

The counterpart to the optimal σ-field $\mathcal{C}_0$ is the "optimal algebra"

$$\mathcal{A}_0 := \{ f \in \mathcal{M} : f \cdot v \in \mathcal{V} \text{ for all } v \in \mathcal{V} \},$$

which by Proposition 1.5(b) is a $\sigma(\mathcal{M}, \mathcal{L})$-closed sub-algebra of $\mathcal{M}$ containing e. The following theorem tells why $\mathcal{A}_0$ is called optimal:

Theorem 4.3:

$f \in \mathcal{M}$ is optimal if and only if $f \in \mathcal{A}_0$ holds.

Proof: (a): let $f \in \mathcal{M}$ be optimal. Putting $\underline{L} = \underline{S}$ we obtain via Proposition 1.5(c)

$$(S * f)(P) \leq [S * (f \pm \tfrac{1}{n} v)](P) = (S * f)(P) \pm \tfrac{2}{n}(f \cdot v)(P) + \tfrac{1}{n^2}(S * v)(P),$$

for all $n \in \mathbb{N}$, thus $\pm(f \cdot v)(P) \geq -\tfrac{1}{2n}(S * v)(P)$ entailing $(f \cdot v)(P) = 0$ whenever $v \in V$ and $P \in \mathcal{P}$. This in turn yields $f \in \mathcal{A}_0$.

(b): assume vice versa that $f \in \mathcal{A}_0$. For $P \in \mathcal{P}$ we choose Π_P as in Theorem 2.4. Then $\{f_-, f_+\} \subseteq \mathcal{A}_0$ implies via Theorem 2.4(a),(b)

$$|f - \Pi_P f|(P) \leq |f_+ - \Pi_P f_+|(P) + |f_- - \Pi_P f_-|(P)$$
$$= f_+(P) - (\Pi_P f_+)(P) + f_-(P) - (\Pi_P f_-)(P) = 0.$$

On the other hand we get by Proposition 2.5 for all $v \in \mathcal{V}$

$$[S*(\Pi_P v)](P) = [(\Pi_P v) \cdot (\Pi_P v)](P)$$
$$= [\Pi_P(v \cdot \Pi_P v)](P) = [v \cdot \Pi_P v](P) = 0 \ .$$

If we substitute in Lemma 4.2 for g subsequently $\Pi_P f - f$ and then $\Pi_P v$, we get with the help of Theorem 2.7

$$(L_P * f)(P) = (L_P * \Pi_P f)(P) = [L_P * (\Pi_P f + \Pi_P v)](P)$$
$$\leq [\Pi_P(L_P * (f + v))](P) = [L_P * (f + v)](P)$$

for every loss function $\underline{L}$, which shows optimality of f. $\qquad\qquad\square$

Remark: As in the proof of Theorem 4.3, it is shown also in [*Torgersen 1980, Theorem 3*] that $f \in \mathcal{M}$ is an element of $\mathcal{A}_0$ if and only if f is $\underline{S}$-optimal. Therefore $\underline{S}$ is a universal loss function even in the generalized sense. $\qquad\qquad\triangle$

Now let us turn towards admissibility:

Definition:

Let $\underline{L} = (L_P)_{P \in \mathcal{P}}$ be a loss function on B (B^*). An element $F \in \mathcal{M}^B$ ($^B\mathcal{M}$) is called "$\underline{L}$-admissible", if for $V \in \mathcal{V}^B$ ($^B\mathcal{V}$) the relation

$$[L_P * (F + V)](P) \leq (L_P * F)(P) \quad \text{for all } P \in \mathcal{P}$$

yields $V = o$.

Theorem 4.4:

Let $\underline{L}$ be a strictly convex loss function. Then every $\underline{L}$-optimal generalized estimator $f \in \mathcal{M}$ is also $\underline{L}$-admissible, and thus uniquely determined among unbiased alternatives $f + v$, where $v \in \mathcal{V}$.

Proof: Assume that $v \in \mathcal{V}$ fulfills $[L_P * (f + v)](P) \leq (L_P * f)(P)$ for all $P \in \mathcal{P}$. Now $\underline{L}$-optimality of f and Proposition 1.4(b) yield – putting $g := f + v$ –

$$(L_P * f)(P) \leq [L_P * (f + \frac{1}{2}v)](P) = [L_P * (\frac{1}{2}f + \frac{1}{2}g)](P)$$
$$\leq \frac{1}{2}(L_P * f)(P) + \frac{1}{2}(L_P * g)(P) \leq (L_P * f)(P).$$

Due to convexity of the functions L_P, we know that

$$\hat{h}(z) := \tfrac{1}{2}L_P(\hat{f}(z)) + \tfrac{1}{2}L_P(\hat{g}(z)) - L_P(\tfrac{1}{2}\hat{f}(z) + \tfrac{1}{2}\hat{g}(z)) \geq 0 \quad \text{holds for all } z \in Z,$$

and hence infer by $\int \hat{h}\, d\hat{P} = 0$ and by strict convexity of L_P that $\hat{f}(z) = \hat{g}(z)$ holds $\hat{P}$-almost surely. This implies

$$[S*(f-g)](P) = \int \left(\hat{f}(z) - \hat{g}(z)\right)^2 \hat{P}(dz) = 0 \quad \text{for all } P \in \mathcal{P},$$

which by Proposition 1.3(b) and Proposition 1.5(c) entails $v = g - f = o$. $\qquad\square$

Example: Denote by $\Omega = [-1,1]$, by $\mathcal{F}$ the system of *Borel* sets in Ω, by $P \in ca(\Omega, \mathcal{F})$ the (suitably normed) *Lebesgue* measure, and by $\mathcal{P} := \{P\}$. Let $L_P(t) := \max\{|t| - 1, 0\}$. For $\gamma := 1_{[-1,0]} - 1_{[0,1]} \in X$, we have $v := \dot\gamma \in \mathcal{V} \setminus \{o\} \subseteq \mathcal{M} \simeq L^\infty(P)$, but for $f := o$ we get

$$0 = [L_P*(f+v)](P) = (L_P*f)(P),$$

whence f is optimal though inadmissible with respect to $\underline{L} = (L_P)$. $\qquad\bowtie$

4.4. The role of sufficiency and completeness

The following results show that optimality, admissibility, and sufficiency are interrelated in the same way as in the classical context. First let us prove that every admissible generalized estimator belongs to the minimal sufficient sub-algebra:

Theorem 4.5:

> Let $\underline{L}$ be an arbitrary loss function. Then every $\underline{L}$-admissible $f \in \mathcal{M}$ is an element of the minimal sufficient sub-algebra $\mathcal{A}_m$.

Proof: If Π denotes the sufficient projection onto $\mathcal{A}_m$ and f is $\underline{L}$-admissible, then for $v := \Pi f - f \in \mathcal{V}$ we have, by Proposition 3.4(a)(v),

$$L_P*(f+v) = L_P*(\Pi f) \leq \Pi(L_P*f),$$

and hence a fortiori

$$[L_P*(f+v)](P) \leq (L_P*f)(P) \quad \text{for all } P \in \mathcal{P},$$

which implies $v = o$ yielding $f = \Pi f \in \mathcal{A}_m$. $\qquad\square$

66

Corollary 4.6:

The relation $\mathcal{A}_0 \subseteq \mathcal{A}_m$ holds true.

Proof: $\underline{L} = \underline{S}$ is a strictly convex loss function. The asserted relation now follows from an application of Theorem 4.3, Proposition 4.4 and Theorem 4.5. $\qquad\square$

In accordance to the classical definition we call a sub-algebra $\mathcal{A}$ of $\mathcal{M}$ "complete", if $\mathcal{A} \cap \mathcal{V} = \{o\}$ holds. The following theorem generalizes both the results by *Rao/Blackwell* and *Lehmann/Scheffé*, as well as their (partial) converse due to *Bahadur* in the classical setup (see section 4.1. above). Here we do not need any further assumptions on the structure of the underlying experiment, an advantage of the transition from classical to generalized estimators.

Theorem 4.7:

The following assertions are equivalent:

(a) $\mathcal{A}_0$ is sufficient;

(b) $\mathcal{A}_m = \mathcal{A}_0$;

(c) $\mathcal{A}_m$ is complete;

(d) there is a complete and sufficient sub-algebra in $\mathcal{M}$;

(e) every (generalized) unbiased estimable parameter ϑ possesses a (generalized) UMVUE.

Proof: (a) $\Rightarrow$ (b) is a consequence of Corollary 4.6. (c) $\Rightarrow$ (d) is trivial.
(b) $\Rightarrow$ (c): $\mathcal{A}_0$ is complete because of $S * v = v \cdot v \in \mathcal{V}_+$ for all $v \in \mathcal{A}_0 \cap \mathcal{V}$, since Proposition 1.3(b) means $\mathcal{V}_+ = \{o\}$.
(d) $\Rightarrow$ (e): let $\mathcal{A}$ be a complete and sufficient sub-algebra and Π the sufficient projection onto $\mathcal{A}$. Since

$$\Pi v \in \mathcal{V} \cap \mathcal{A} = \{o\} \text{ for all } v \in \mathcal{V},$$

Proposition 3.4(a) entails

$$(v \cdot \Pi f)(P) = [\Pi(v \cdot \Pi f)](P) = [(\Pi v) \cdot (\Pi f)](P) = 0 \quad \text{for all } P \in \mathcal{P},$$

so that $f^\star := \Pi f \in \mathcal{A}_0$, as well as $f^\star - f \in \mathcal{V}$. By Theorem 4.3, the element $f^\star$ is $\underline{S}$-optimal.

(e) $\Rightarrow$ (a): for $f \in \mathcal{M}$ we denote by $\Pi_0 f = f^\star$ the $\underline{S}$-optimal element fulfilling $f - f^\star \in \mathcal{V}$. Due to Proposition 4.4, the element $f^\star$ is uniquely determined. Now Theorem 4.3 implies $\Pi_0(\mathcal{M}) \subseteq \mathcal{A}_0$. Since $h = \Pi_0 h \in \mathcal{V}$ holds for all $h \in \mathcal{M}$ we conclude

$$\Pi_0 g = g \quad \text{and} \quad \Pi_0(f \cdot g) = g \cdot \Pi_0 f \quad \text{for all } f \in \mathcal{M} \text{ and all } g \in \mathcal{A}_0,$$

because $g \cdot \Pi_0 f \in \mathcal{A}_0$ and the relation

$$\Pi_0(f \cdot g) - g \cdot \Pi_0 f = \Pi_0(f \cdot g) - f \cdot g + (f - \Pi_0 f) \cdot g \in \mathcal{V}$$

pertain. Thus Proposition 3.4(b) yields sufficiency of $\mathcal{A}_0$. $\qquad\square$

Remark: Corollary 4.6 as well as Theorem 4.7 are immediate consequences of Theorems 12 and 14 in [*Torgersen* 1980], in which paper the author chose a different approach to define $\mathcal{M}$ (see appendix). From the proof of Theorem 4.7, (d) $\Rightarrow$ (e) we see that every sub-algebra of $\mathcal{M}$ which is sufficient and complete, already coincides with $\mathcal{A}_0$. $\qquad\triangle$

4.5. Existence of nontrivial optimal estimators

To generalize the criterion for existence of non-trivial optimal estimators mentioned in section 4.1., we need a notion analoguous to the Q-support $[\mathcal{H}]_Q$ of a sub-set $\mathcal{H} \subseteq L^1(Q)$:

Definition:

Let $\mathcal{H} \subseteq \mathcal{M}$. An idempotent $t = t \cdot t \in \mathcal{M}$ is called "support of $\mathcal{H}$", if

(i) $t \cdot f = f$ for all $f \in \mathcal{H}$ holds as well as

(ii) $t \cdot g = o$ for all $g \in \mathcal{M}$ fulfilling $g \cdot f = o$ for all $f \in \mathcal{H}$.

Proposition 4.8:

For every $\mathcal{H} \subseteq \mathcal{M}$ there is one and only one support of $\mathcal{H}$ which we denote by $t_\mathcal{H}$. Furthermore, $\mathcal{H} \subseteq \mathcal{K} \subseteq \mathcal{M}$ yields $t_\mathcal{H} \leq t_\mathcal{K}$.

68

Proof: Assume that t and s are supports of $\mathcal{H}$ and $\mathcal{K}$, respectively. Then

$$(e - s) \cdot \mathcal{H} \subseteq (e - s) \cdot \mathcal{K} = \{o\}$$

yields $t(e - s) = o$ entailing $t = t \cdot s \leq e \cdot s = s$, since by $o \leq S * (e - t) = e - t$ we have $o \leq S * t = t \leq t + (e - t) \cdot (e - t) = e$. This implies uniqueness as well as the inequality claimed. To show existence we note that for $f \in \mathcal{M}$ the element

$$t_{\{f\}} := \sup_{n \in \mathbb{N}} \left[\inf\{e, n|f|\} \right] \in \mathcal{M}$$

fulfills $\hat{t}_{\{f\}} = 1_{U(f)}$, where we denote by $U(f)$ the closed support of $\hat{f}$,

$$U(f) := \overline{\{z \in Z : \hat{f}(z) \neq 0\}} \in \mathcal{B}_{\mathcal{M}}$$

which is clopen because Z is extremally disconnected (*Vulikh* [1967, p.99] calls $t_{\{f\}}$ "trace of f"). Now we define

$$t_{\mathcal{H}} := \sup_{f \in \mathcal{H}} t_{\{f\}} \in \mathcal{M}.$$

Then $\hat{t}_{\mathcal{H}} = 1_{\overline{U}}$, where we have put

$$U := \bigcup_{f \in \mathcal{H}} U(f)$$

(clearly U is open in Z, hence $\overline{U}$ is clopen). Evidently we get

$$t_{\mathcal{H}} \cdot t_{\mathcal{H}} = t_{\mathcal{H}} \text{ and } t_{\mathcal{H}} \cdot f = f \text{ for all } f \in \mathcal{H}.$$

If $g \in \mathcal{M}$ fulfills $g \cdot f = o$ for all $f \in \mathcal{H}$ and $z \in \overline{U}$ then there is a net $(z_i)_{i \in I}$ in Z and $f_i \in \mathcal{H}$ with $z_i \in U(f_i)$ for all $i \in I$, such that $z_i \to z$ as $i \in I$. Now $g \cdot f_i = o$ implies $\hat{g}(z_i) = 0$ for all $i \in I$ yielding $\hat{g}(z) = 0$, which means $t_{\mathcal{H}} \cdot g = o$. $\qquad\square$

Remark: If $\mathcal{E}$ is coherent, then the relation $t_{\mathcal{H}} \cdot t_{\mathcal{H}} = t_{\mathcal{H}} = \dot{\varphi} \in \dot{X}$ and Proposition 1.5(d) together entail $t_{\mathcal{H}} = \dot{1}_A$, where $A := \varphi^{-1}(\{1\}) \in \mathcal{F}$. Then for all $P \in \mathcal{P}$ we obtain

$$P(A \setminus \{\omega \in \Omega : \varphi(\omega) \neq 0\}) = 0 \quad \text{for all } \dot{\varphi} \in \mathcal{H}$$

as well as

$$P^\star\Big(\bigcup_{\dot{\varphi} \in \mathcal{H}} \{\omega \in \Omega : \varphi(\omega) \neq 0\} \setminus A \Big) = 0,$$

where $P^\star$ denotes the outer measure on 2^Ω corresponding to P. In particular for the dominated case $\mathcal{P} \cong Q$, we therefore have $t_{\mathcal{H}} = (1_{[\mathcal{H}]_Q})^{\cdot}$ (cf. [*Bomze* 1986]). $\qquad\triangle$

Lemma 4.9:

If $s = s \cdot s \in \mathcal{M}$ and $t_V \leq s$, then $s \in \mathcal{A}_0$.

Proof: The inequality $s \leq e$ and Proposition 1.5(a) yield

$$s \cdot t_V \leq e \cdot t_V = t_V = t_V \cdot t_V \leq s \cdot t_V$$

which means $s \cdot t_V = t_V$. For $v \in V$ we thus infer $s \cdot v = s \cdot t_V \cdot v = t_V \cdot v = v \in V$, implying a fortiori $s \in \mathcal{A}_0$. $\qquad\square$

Definition:

V is called "reducible", if there are non-trivial, $\sigma(\mathcal{M}, \mathcal{L})$-closed linear sub-spaces V_1 and V_2 of V fulfilling

$$V_1 \oplus V_2 = V \quad \text{and} \quad V_1 \cdot V_2 = \{o\};$$

otherwise V is said to be "irreducible".

Theorem 4.10:

(a) Assume that $t_V = e$ holds. Then

$$\mathcal{A}_0 = \mathbb{R} \cdot e \quad \text{if and only if} \quad V \text{ is irreducible.}$$

(b) Assume that $o < t_V < e$ holds. Then $\mathbb{R} \cdot e \neq \mathcal{A}_0$. Furthermore we have

$$\{s \in \mathcal{A}_0 : o < s = s \cdot s < t_V\} = \emptyset \quad \text{if and only if} \quad V \text{ is irreducible.}$$

(c) The following assertions are equivalent:
    (i) $t_V = o$;
    (ii) $V = \{o\}$;
    (iii) $\mathcal{A}_0 = \mathcal{M}$.

Proof: (c) follows from $\mathcal{A}_0 \cap V = \{o\}$.

(b): according to Lemma 4.9 we have $t_V \in \mathcal{A}_0 \setminus \{o, e\}$ and thus, by idempotence of t_V, we get $t_V \in \mathcal{A}_0 \setminus \mathbb{R} \cdot e$. If $o < s < t_V$ and $s \cdot s = s \in \mathcal{A}_0$ hold, then

$$V_1 := s \cdot V \quad \text{and} \quad V_2 := (t_V - s) \cdot V$$

are two $\sigma(\mathcal{M}, \mathcal{L})$-closed, linear sub-spaces of V. Indeed, if

$$s \cdot v_i \to w \text{ w.r.t. } \sigma(\mathcal{M}, \mathcal{L}) \text{ as } i \in I,$$

70

where $v_i \in \mathcal{V}$ for all $i \in I$, then $s \cdot v_i \in \mathcal{V}$ for all $i \in I$ yields $w \in \mathcal{V}$. Moreover, Proposition 1.5(b) entails $w = s \cdot w \in \mathcal{V}_1$. Now $\mathcal{V}_1 \neq \{o\}$, since otherwise we obtained $o = s \cdot t_\mathcal{V} = s$, the last equality being due to $s < t_\mathcal{V}$. Similarly we conclude that $\mathcal{V}_2 \neq \{o\}$. Furthermore, observe that

$$s \cdot (t_\mathcal{V} - s) = s \cdot t_\mathcal{V} - s = o$$

entails $\mathcal{V}_1 \cdot \mathcal{V}_2 = \{o\}$ as well as

$$s \cdot v + (t_\mathcal{V} - s) \cdot v = t_\mathcal{V} \cdot v = v \quad \text{for all } v \in \mathcal{V}$$

implies $\mathcal{V}_1 \oplus \mathcal{V}_2 = \mathcal{V}$. Hence $\mathcal{V}$ is reducible.

Conversely, a non-trivial decomposition of the form

$$\mathcal{V} = \mathcal{V}_1 \oplus \mathcal{V}_2 \quad \text{with} \quad \mathcal{V}_1 \cdot \mathcal{V}_2 = \{o\}$$

gives rise to an idempotent element $s := t_{\mathcal{V}_1} = s \cdot s \in \mathcal{M}$ fulfilling

$$s \cdot \mathcal{V}_1 = \mathcal{V}_1 \subseteq \mathcal{V} \quad \text{and} \quad s \cdot \mathcal{V}_2 = \{o\} \subseteq \mathcal{V},$$

which implies $s \cdot \mathcal{V} \subseteq \mathcal{V}$, whence we get $s \in \mathcal{A}_0$. Now $\mathcal{V}_1 \subseteq \mathcal{V}$ via Proposition 4.8 entails $s \leq t_\mathcal{V}$, by $\mathcal{V}_1 \neq \{o\}$ and $\mathcal{V}_2 \neq \{o\}$ we infer $s > o$ and $s < t_\mathcal{V}$.

(a): we know by Proposition 2.1 that $\mathbb{R} \cdot e \neq \mathcal{A}_0$ holds if and only if there exists an idempotent $s \in \mathcal{A}_0$ fulfilling $o < s < e = t_\mathcal{V}$. The remainder is completely analoguous to the proof of the second assertion in (b). $\qquad\square$

4.6. Generalized Banach space valued unbiased estimation

Next we shall deal with optimality and admissibility of B-valued generalized estimators: from a loss function $\underline{L} = (L_P)_{P \in \mathcal{P}}$ on $\mathbb{R}$ and a linear functional $y \in B$ we obtain with the help of Proposition 1.13 a loss function $\underline{L}^y := (L_P^y)_{P \in \mathcal{P}}$ on B of particular type. Similarly, every evaluation functional $x \in B \subseteq B^{**}$ gives rise to a loss function $\underline{L}^x := (L_P^x)_{P \in \mathcal{P}}$ on B^*. Both loss functions are in an evident way related to $\underline{L}$ with respect to optimality and admissibility:

Proposition 4.11:

Let $\underline{L}$ be a loss function on $\mathbb{R}$ and $(x,y) \in B \times B^*$. Then

(a) $F \in \mathcal{M}^B$ is $\underline{L}^y$-optimal if and only if $\langle F, y \rangle$ is $\underline{L}$-optimal;
$F \in {}^B\mathcal{M}$ is $\underline{L}^x$-optimal if and only if $\langle x, F \rangle$ is $\underline{L}$-optimal;

(b) $F \in \mathcal{M}^B$ is $\underline{L}^y$-admissible if and only if $\langle F, y \rangle$ is $\underline{L}$-admissible;
$F \in {}^B\mathcal{M}$ is $\underline{L}^x$-admissible if and only if $\langle x, F \rangle$ is $\underline{L}$-admissible.

Proof: Let $F \in \mathcal{M}^B$ and $y \in B^*$. We may and do assume without loss of generality that $y \neq o$, so that $\langle x, y \rangle = 1$ holds for some element $x \in B$. Hence for $v \in V$ we have $G := v \cdot x \in V^B$ and $\langle G, y \rangle = v$. On the other hand, $\langle G, y \rangle \in V$ holds for all $G \in V^B$. The assertions in (a) and (b) now are easily derived from Propositions 1.12(d) and 1.13. The proof for $F \in {}^B\mathcal{M}$ follows the same lines. $\square$

As an immediate consequence we see (Proposition 4.12 below) that if $\underline{L}$ is a strictly convex loss function, then $\underline{L}^x$-optimality already yields $\underline{L}^x$-admissibility. Since $\underline{L}^x$ for $B \neq \mathbb{R}$ is not strictly convex, this implication is special because in general it holds true only for strictly convex loss functions on B^* (cf. example following Proposition 4.4).

Proposition 4.12:

Let $\underline{L}$ be a strictly convex loss function on B^*. Then
every $\underline{L}$-optimal $F \in {}^B\mathcal{M}$ is also $\underline{L}$-admissible.

Proof: By analogy to the case $B^* = \mathbb{R}$ in Proposition 4.4 we deduce via Theorem 1.14 from the inequality

$$[L_{P^*}(F + H)](P) \leq (L_{P^*}F)(P) \quad \text{for all } H \in {}^B V$$

that $\hat{H}(z) = o$ holds $\hat{P}$-almost surely, which in turn entails for any $x \in B$ the relation

$$[S_*\langle x, H \rangle](P) = \int (\langle x, \hat{H}(z) \rangle)^2 \, \hat{P}(dz)$$

$$\leq \int (|||\hat{H}(z)||| \; |||x|||)^2 \, \hat{P}(dz) = 0 \quad \text{for all } P \in \mathcal{P},$$

whence $\langle x, H \rangle = o$ follows for all $x \in B$, yielding $H = o$. $\square$

Theorem 4.13:

Let $\underline{L}$ be a loss function on B. Then
every $\underline{L}$-admissible $F \in \mathcal{M}^B$ is an element of $\mathcal{A}_m{}^B$.

72

Proof: In complete analogy to the proof of Theorem 4.5 the claimed assertion follows from Proposition 3.4(a)(v), because Proposition 1.9 and Proposition 3.4(a)(ii) yield the relation $\Pi^B F - F \in \mathcal{V}^B$, where Π denotes the ($\sigma(\mathcal{M},\mathcal{L})$-$\sigma(\mathcal{M},\mathcal{L})$-continuous) sufficient projection onto $\mathcal{A}_m$. $\qquad\square$

Remark: If we replace in the arguments for Proposition 4.12 and Theorem 4.13 (i) the dual space B^* by an arbitrary *Banach* space B, (ii) the map $\hat{F} \in C^B(Z)$ by an $\mathcal{F}$-$\|\|\cdot\|\|$-*Borel*–measurable function $\Phi : \Omega \to B$, and (iii) the experiment $\hat{\mathcal{E}}$ by the experiment $\mathcal{E}$, then we obtain, without assuming separability of the dual space B^*, the classical result that every $\underline{L}$-admissible estimator Φ is $\mathcal{C}_m$-$\sigma(B,B^*)$-*Borel*–measurable, if $\underline{L}$ is a strictly convex loss function on B, provided Φ is *Bochner* integrable with respect to all $P \in \mathcal{P}$ and has finite risk $\int L_P \circ \Phi \, dP < +\infty$: indeed, proceeding analogously we only have to use sufficiency of the sub- σ-fields $\mathcal{C}(\supseteq \mathcal{C}_m)$ and *Jensen's* inequality (see, e.g. [*Bomze* 1984]). For separable B such an estimator Φ is thus even $\mathcal{C}_m$-$\|\|\cdot\|\|$-*Borel*–measurable (cf. Theorem 1 in [*Kozek* 1980] or Theorem 2 in [*Kozek/Suchanecki* 1980], where the proof seems to be rather technical for our purposes). $\qquad\triangle$

Next we show that $F \in {}^B\mathcal{M}$ is optimal if and only if evaluations $\langle x, F \rangle \in \mathcal{M}$ of F at x are optimal for all $x \in B$:

Theorem 4.14:
   Let $F \in {}^B\mathcal{M}$. Then the following assertions are equivalent:

(a) F is optimal;

(b) F is $\underline{S}^x$-optimal for all $x \in B$;

(c) $F \in {}^B\mathcal{A}_0$.

Proof: (a) $\Rightarrow$ (b) is trivial, (c) and (b) are equivalent due to Theorem 4.3 and Proposition 4.11(a).
(c) $\Rightarrow$ (a): let $F \in {}^B\mathcal{A}_0$, $H \in {}^B\mathcal{V}$, $P \in \mathcal{P}$, and $\underline{L}$ be a loss function on B^*. By analogy to the proof of Theorem 4.3, we deduce from Proposition 1.9 and from the relation $\langle x, F \rangle \in \mathcal{A}_0$ the equalities

$$|\langle x, {}^B(\Pi_P)F - F \rangle|(P) = |\Pi_P \langle x, F \rangle - \langle x, F \rangle|(P) = 0,$$

and from $\langle x, F \rangle \in \mathcal{V}$ via Lemma 4.2 the equality $|\Pi_P \langle x, H \rangle|(P) = 0$. These facts imply for any measures $\mu_i \in \mathcal{L}_+$, $1 \leq i \leq n$, fulfilling $\sum_{i=1}^{n} \mu_i = P$, that

$$|\langle x, {}^B(\Pi_P)(F + H) \rangle - \langle x, F \rangle|(\mu_i) \leq |\langle x, {}^B(\Pi_P)F - F \rangle|(\mu_i) + |\Pi_P(\langle x, H \rangle)|(\mu_i) = 0$$

holds true for all $y \in B^*$, which entails

$$\sum_{i=1}^{n} [\langle x_i, {}^{B}(\Pi_P)(F+H)\rangle + \beta_i e](\mu_i) = \sum_{i=1}^{n} [\langle x_i, F\rangle + \beta_i e](\mu_i).$$

Again by Proposition 1.11(c), we derive the equality

$$(L_P * F)(P) = [L_P * {}^{B}(\Pi_P)(F+H)](P)$$
$$\leq \Pi_P[L_P * (F+H)](P) = (L_P * (F+H))(P)$$

(see Theorems 2.7 and 2.4(a)), which proves the last implication. $\qquad\square$

Remark: For reflexive Banach spaces $B = B^{**}$ we therefore have a characterization of optimal generalized estimators $F \in \mathcal{M}^B = {}^{B^*}\mathcal{M}$ which is analogous to Theorem 4.3. Even for general Banach spaces B, this characterization has a certain meaning in that one may view ${}^{B^*}\mathcal{M}$ as completion of $\dot{X}^B$ (cf. Proposition 1.8), as well as one may extend convex loss functions $\underline{L} = (L_P)_{P\in\mathcal{P}}$ on B to convex loss functions $\overline{\underline{L}} = (\overline{L_P})_{P\in\mathcal{P}}$ on $B^{**} = (B^*)^*$ as in Proposition 1.11(e). Furthermore we easily deduce from Proposition 4.11(a) that an element $F \in \mathcal{M}^B$ is $\underline{S}^y$-optimal for all $y \in B^*$ if and only if F belongs to $\mathcal{A}_0{}^B$. If we moreover know that there exist $\sigma(\mathcal{M},\mathcal{L})$-$\sigma(\mathcal{M},\mathcal{L})$-continuous maps $\Pi_P : \mathcal{M} \to \mathcal{A}_0$ having the properties described in Theorem 2.4 (this in turn is, for instance, guaranteed by the equality $\mathcal{A}_0 = \mathcal{A}_m$, putting $\Pi_P = \Pi$ for all $P \in \mathcal{P}$, where Π is the sufficient projection onto $\mathcal{A}_m$; one may find sufficient conditions for $\mathcal{A}_0 = \mathcal{A}_m$ to hold in chapter 5), we are able to derive optimality of F employing Proposition 1.9(b). This may be achieved by analogy to the proof of (c) $\Rightarrow$ (a) in Theorem 4.14, using the fact that F belongs to $\mathcal{A}_0{}^B$ (replace ${}^{B}(\Pi_P)$ by $(\Pi_P)^B$). Note that in the (trivial) case $\mathcal{A}_0 = \mathbb{R} \cdot e$ even the conclusion (c) $\Rightarrow$ (a) requires no additional assumptions: $\qquad\triangle$

Lemma 4.15:

The following assertions are equivalent:
(a) $\mathcal{A}_0 = \mathbb{R} \cdot e$;
(b) $\mathcal{A}_0{}^B = B \cdot e$;
(c) ${}^{B}\mathcal{A}_0 = B^* \cdot e$.

Proof: (a) $\Rightarrow$ (b): if $F \in \mathcal{A}_0{}^B \setminus (B \cdot e)$, then there is a functional $y \in B^*$ fulfilling $\langle F, y \rangle \in \mathcal{A}_0 \setminus \mathbb{R} \cdot e$, since otherwise for all $y \in B^*$ there existed a number $\alpha_y \in \mathbb{R}$ with $\langle F, y \rangle = \alpha_y \cdot e$. Fix a probability measure $P \in \mathcal{P}$. Because $\alpha_y = \langle F(P), y \rangle$ holds for all $y \in B^*$, the functional $y \mapsto \alpha_y$ from B^* to $\mathbb{R}$ were linear and $\sigma(B^*, B)$-continuous, and thus

of the type $\alpha_y = \langle x, y \rangle$ for a suitably chosen $x \in B$, entailing the contradiction $F = x \cdot e$.
Similarly one proves (a) $\Rightarrow$ (c).

(c)$\Rightarrow$(a): for $f \in \mathcal{A}_0$ choose $(x, y) \in B \times B^*$ fulfilling $\langle x, y \rangle = 1$ and $y' \in B^*$ such that $f \cdot (y \cdot e) = y' \cdot e$ holds. Hence we have

$$f = \langle x, y \rangle f = \langle x, f \cdot (y \cdot e) \rangle = \langle x, y' \rangle e \in \mathbb{R} \cdot e.$$

By analogy, the proof of the implication (b) $\Rightarrow$ (a) is obtained. $\qquad\square$

As an immediate consequence we now obtain the multivariate counterpart of Theorem 4.10, the remark after Theorem 4.14 mutatis mutandis being also valid:

Theorem 4.16:

(a) Assume that $\mathcal{V}$ has full support, which means that $t_\mathcal{V} = e$ holds. Then the following assertions are equivalent:

    (i) $\mathcal{V}$ is irreducible;

    (ii) $\mathcal{A}_0{}^B = B \cdot e$;

    (iii) ${}^B\mathcal{A}_0 = B^* \cdot e$;

    (iv) every optimal $F \in {}^B\mathcal{M}$ is trivial.

In this case there is even no non-trivial optimal $F \in \mathcal{M}^B$.

(b) If $o < t_\mathcal{V} < e$, then $B \cdot e \neq \mathcal{A}_0{}^B$ and $B^* \cdot e \neq {}^B\mathcal{A}_0$. Furthermore the following assertions are equivalent:

    (i) $\mathcal{V}$ is irreducible;

    (ii) $\{s \cdot (x \cdot e) : s \in \mathcal{A}_0, o < s = s \cdot s < t_\mathcal{V}\} = \emptyset$ for all $x \in B$;

    (iii) $\{s \cdot (y \cdot e) : s \in \mathcal{A}_0, o < s = s \cdot s < t_\mathcal{V}\} = \emptyset$ for all $y \in B^*$.

(c) The following assertions are equivalent:

    (i) $t_\mathcal{V} = o$;

    (ii) $\mathcal{A}_0{}^B = \mathcal{M}^B$;

    (iii) ${}^B\mathcal{A}_0 = {}^B\mathcal{M}$.

Proof: follows from Lemma 4.15, Theorems 4.10, and 4.14. $\qquad\square$

5. Invariance

This chapter treats invariance under arbitrary groups G of transitions. A "transition" $T : \mathcal{L} \to \mathcal{L}$ is a linear, isometric lattice isomorphism on $\mathcal{L}$. This concept generalizes model preserving transformations acting on the sample space Ω. After investigating some important properties of transitions in section 5.1., we shall show in section 5.2., that every element of $\mathcal{A}_m$, particularly every generalized estimator that is admissible with respect to a strictly convex loss function, already is invariant under the adjoint $T^* : \mathcal{M} \to \mathcal{M}$ of a transition $T : \mathcal{L} \to \mathcal{L}$. This generalizes the result in the classical context that the minimal σ-field $\mathcal{C}_m$ is contained in the σ-field of the almost invariant sets (see [*Basu* 1970]). Moreover we propose a criterion for the case that, conversely, every invariant generalized estimator already is optimal. In the sequel it will turn out that this happens indeed for the so-called full model.

In section 5.3., we shall exhibit for a large class of groups an explicit representation of the sufficient projection onto the algebra of invariant generalized estimators which enables us to calculate optimal generalized estimators in certain situations important for application. Thereupon we treat in section 5.4. a special case which is in close correspondence to the procedure proposed in [*Basu* 1970], namely to seek for a minimal sufficient statistic which in addition is a maximal invariant under G: if – roughly spoken – we can parametrize $\mathcal{P}$ by a maximal invariant χ defined on Ω such that every $P \in \mathcal{P}$ is concentrated on a level set of χ, we are able to show that optimality and invariance are equivalent notions in this case. We close this chapter by transferring, in section 5.5., the results indicated above to invariant estimation in *Banach* spaces.

5.1. Transitions

Let G be a group and $T_\theta : \mathcal{L} \to \mathcal{L}$, for $\theta \in G$, be a transition on $\mathcal{L}$ fulfilling $T_\theta P = P$ for all $P \in \mathcal{P}$. We furthermore assume that $T_\theta \circ T_\tau = T_{\theta\tau}$ holds for $\{\theta, \tau\} \subseteq G$. Evidently the adjoint map $T_\theta^* : \mathcal{M} \to \mathcal{M}$ of T_θ is $\sigma(\mathcal{M}, \mathcal{L})$-$\sigma(\mathcal{M}, \mathcal{L})$-continuous.

Remark: Let θ be a "transformation" of the experiment $\mathcal{E}$, i.e. an invertible map from Ω onto itself which is bimeasurable with respect to $\mathcal{F}$, such that $\{\tilde{T}_\theta P : P \in \mathcal{P}\} \subseteq \mathcal{P}$ holds, denoting for $\mu \in ca(\Omega, \mathcal{F})$

$$\tilde{T}_\theta \mu(A) := \mu(\theta^{-1}(A)) \quad \text{for all } A \in \mathcal{F}.$$

This transformation gives rise to an invertible map $\tilde{T}_\theta : ca(\Omega, \mathcal{F}) \to ca(\Omega, \mathcal{F})$ which in both directions preserves order. Now from

$$\sup\{\tilde{T}_\theta \mu, \tilde{T}_\theta \nu\} = \tilde{T}_\theta(\sup\{\mu, \nu\}) \quad \text{for all } \{\mu, \nu\} \subseteq \mathcal{L}$$

we on one hand deduce

$$\|\tilde{T}_\theta \mu\| = |\tilde{T}_\theta \mu|(\Omega) = [\tilde{T}_\theta|\mu|](\Omega) = |\mu|(\Omega) = \|\mu\|$$

and on the other hand obtain

$$\tilde{T}_\theta(\mathcal{L}) \subseteq \mathcal{L}.$$

Indeed, for $\mu \in \mathcal{L}$ and $\sigma \in ca(\Omega, \mathcal{F})$ with $\sigma \perp \mathcal{P}$ we also have $\tilde{T}_\theta^{-1}\sigma \perp \mathcal{P}$, entailing by Proposition 1.1(c) $\mu \perp \tilde{T}_\theta^{-1}\sigma$ and thus $\tilde{T}_\theta\mu \perp \sigma$, which in turn yields $\tilde{T}_\theta\mu \in \mathcal{L}$, again via Proposition 1.1(c). By consequence, the restricted map

$$T_\theta := \tilde{T}_\theta|_{\mathcal{L}}$$

is a transition on $\mathcal{L}$. Furthermore, the transformation formula for integrals applied to an arbitrary function $\varphi \in X$ yields $T_\theta^*\dot\varphi = (\varphi \circ \theta)^{\boldsymbol{\cdot}}$ and hence $T_\theta^*(\dot{X}) \subseteq \dot{X}$. If, in addition, θ is "model preserving", which means that $T_\theta P = P$ holds for all $P \in \mathcal{P}$, then we face the situation described in the preceding paragraph. The relation $\dot\varphi = T_\theta^*\dot\varphi$, for some $\varphi \in X$, implies the existence of a θ-invariant function $\overline\varphi \in X$ fulfilling $\dot{\overline\varphi} = \dot\varphi$ (cf. [*Lehmann* 1959, p.225]). Note that the group $G := \{\theta^n : n \in \mathbb{Z}\}$, the system $\mathcal{G} := 2^G$, and the counting measure $\tilde\eta$ on $\mathcal{G}$, satisfy the conditions specified in the remark following Proposition 5.13 below. By $\tilde\theta : A \mapsto \theta^{-1}(A)$ a map on $\mathcal{F}$ is defined that evidently preserves intersection, union, and complementation. Since $\tilde\theta(\mathcal{N}) \subseteq \mathcal{N}$, one may extend $\tilde\theta$ in a natural way to a map defined on the factor $\mathcal{F}/\mathcal{N}$. If, conversely, $T : \mathcal{L} \to \mathcal{L}$ is a transition in the general setup described at the beginning, which fulfills $T^*(\dot{X}) \subseteq \dot{X}$, then Lemma 5.1 below implies that T is induced – via the formula $T\mu = \mu \circ \tilde\theta$ – by a (set-)algebraic isomorphism $\tilde\theta$ on $\mathcal{F}/\mathcal{N}$ in the sense used above. Indeed, for $\varphi \in X$ we easily see that $\dot\varphi \cdot \dot\varphi = \dot\varphi$ if and only if $\dot\varphi = 1_A$ for a suitably chosen set $A \in \mathcal{F}$, an obvious choice being $A := \varphi^{-1}(\{1\})$. $\qquad \triangle$

Lemma 5.1:

Let $T : \mathcal{L} \to \mathcal{L}$ be a transition. Then

(a) for convex $L : \mathbb{R} \to \mathbb{R}$ and $f \in \mathcal{M}$ we have $L*(T^*f) = T^*(L*f)$;

(b) T^* is multiplicative and hence an algebraic isomorphism on $\mathcal{M}$;

(c) there is a homeomorphism $\hat{T} : Z \to Z$ fulfilling $\widehat{T^*f} = \hat{f} \circ \hat{T}$ for all $f \in \mathcal{M}$.

Proof: (a): Since

$$\|(T\mu)_+\| - \|(T\mu)_-\| = \|\mu_+\| - \|\mu_-\| \quad \text{for all } \mu \in \mathcal{L},$$

we have $T^*e = e$ and hence

$$
\begin{aligned}
L*(T^*f) &= \sup_{n \in \mathbb{N}}[\alpha_n T^* f + \beta_n e]\\
&= \sup_{n \in \mathbb{N}} T^*(\alpha_n f + \beta_n e)\\
&= T^*(\sup_{n \in \mathbb{N}}[\alpha_n f + \beta_n e]) = T^*(L*f)
\end{aligned}
$$

for any convex function $L : \mathbb{R} \to \mathbb{R}$, because T^* and $(T^*)^{-1}$ both are linear, positive maps defined on $\mathcal{M}$.

(b) is a consequence of

$$T^*(f \cdot g) = \frac{1}{2}[S*T^*(f+g) - S*(T^*f) - S*(T^*g)] = (T^*f) \cdot (T^*g).$$

(c) is proved, in [*Dunford/Schwartz 1964, p.278*]. $\qquad\square$

By means of these observations we now are able to prove the following ergodic theorem, which serves as an auxiliary result:

Lemma 5.2:

Let $T^* : \mathcal{M} \to \mathcal{M}$ be an algebraic isomorphism which is $\sigma(\mathcal{M}, \mathcal{L})$-$\sigma(\mathcal{M}, \mathcal{L})$-continuous and thus the adjoint of a linear map $T : \mathcal{L} \to \mathcal{L}$. Assume that $TP = P$ holds for all $P \in \mathcal{P}$. Then

for all $f \in \mathcal{M}$ there is an element $\overline{f} \in \mathcal{M}$ fulfilling $T^*\overline{f} = \overline{f}$ and

$$\frac{1}{n}\sum_{k=0}^{n-1} T^{*k}f \to \overline{f} \quad \text{w.r.t. } \sigma(\mathcal{M}, \mathcal{L}) \text{ as } n \to \infty.$$

78

Proof: Let $\hat{T} : Z \to Z$ be as in Lemma 5.1. For $U \in \mathcal{B}_\mathcal{M}$ we have $\hat{P}(\hat{T}^{-1}(U)) = (T^* s_U)(P) = s_U(P) = \hat{P}(U)$. Since $\mathcal{M}' := \{U \in \hat{\mathcal{M}} : \hat{P}(\hat{T}^{-1}(U)) = \hat{P}(U)\}$ is a σ-field containing $\mathcal{B}_\mathcal{M}$, and therefore coincides with $\hat{\mathcal{M}}$, we hence obtain

$$\hat{P}(\hat{T}^{-1}(U)) = \hat{P}(U) \quad \text{for all } U \in \hat{\mathcal{M}}.$$

According to Theorem 2.2(a) we have

$$N_f := \{z \in Z : \liminf_{n \in \mathbb{N}} \frac{1}{n} \sum_{k=0}^{n-1} \hat{f} \circ \hat{T}^k(z) < \limsup_{n \in \mathbb{N}} \frac{1}{n} \sum_{k=0}^{n-1} \hat{f} \circ \hat{T}^k(z)\} \in \hat{\mathcal{M}}$$

and evidently $N_f = \hat{T}^{-1}(N_f)$. Now *Birkhoff*'s pointwise ergodic theorem (see, e.g. [*Neveu* 1968, p.247]) applied to $(Z, \hat{\mathcal{M}}, \hat{P})$ yields $\hat{P}(N_f) = 0$ for all $P \in \mathcal{P}$. By defining

$$r(z) := \begin{cases} \lim_{n \to \infty} \frac{1}{n} \sum_{k=0}^{n-1} \hat{f} \circ \hat{T}^k(z), & \text{if } z \in Z \setminus N_f, \\ 0, & \text{otherwise,} \end{cases}$$

we obtain a $\hat{\mathcal{M}}$-measurable function $r : Z \to [-\|f\|, \|f\|]$. Again, by Theorem 2.2 we know that there exist an element $\overline{f} \in \mathcal{M}$ and a set $U \in \mathcal{B}_\mathcal{M}$ fulfilling

$$\hat{\overline{f}}(z) = r(z) \ \hat{\mu}\text{-almost surely, and } \hat{\mu}(U \triangle N_f) = 0 \quad \text{for all } \mu \in \mathcal{L}.$$

Now $s_U \in \mathcal{M}_+$ fulfills $s_U(P) = \hat{P}(N_f) = 0$ for all $P \in \mathcal{P}$, whence we via Proposition 1.3(b) infer $s_U = o$, or, equivalently, $U = \varnothing$, which means

$$\hat{\mu}(N_f) = 0 \quad \text{for all } \mu \in \mathcal{L}.$$

The dominated convergence theorem applied to $(Z, \hat{\mathcal{M}}, \hat{\mu})$ now together with Lemma 5.1(c) entails

$$\frac{1}{n} \sum_{k=0}^{n-1} T^{*k} f \to \overline{f} \text{ w.r.t. } \sigma(\mathcal{M}, \mathcal{L}) \text{ as } n \to \infty$$

as well as $T^* \overline{f} = \overline{f}$ due to $\sigma(\mathcal{M}, \mathcal{L})$-$\sigma(\mathcal{M}, \mathcal{L})$-continuity of T^*. $\qquad\square$

Remark: Lemma 5.2 may be deduced from general ergodic theory as follows: consider any linear, positive contraction which fixes $\mathcal{P}$, that is a linear, positive map $T : \mathcal{L} \to \mathcal{L}$ with $\|T\mu\| \leq \|\mu\|$, all $\mu \in \mathcal{L}$, and $TP = P$, all $P \in \mathcal{P}$. Now fix a measure $\mu \in \mathcal{L}'$. Then the *Cesaro* means $\mu_n := \frac{1}{n} \sum_{k=0}^{n-1} T^k \mu$ fulfill $\|\mu_n - \overline{\mu}\| \to 0$ as $n \to \infty$, where $\overline{\mu} \in \tilde{\mathcal{L}}$ satisfies $T\overline{\mu} = \overline{\mu}$. Indeed, choose some $\alpha_i > 0$ and some $P_i \in \mathcal{P}$, $1 \leq i \leq m$, with $|\mu| \leq Q := \sum_{i=1}^m \alpha_i P_i$, and consider the space

$$\tilde{\mathcal{L}} := \overline{\{\nu \in \mathcal{L} : |\nu| \leq \alpha Q \text{ for some } \alpha \geq 0\}},$$

that is the $\|.\|$-closure of the linear span of all measures ν which are smaller than Q w.r.t. the ordering $\leq$ in $\mathcal{L}$. Then Q is a quasi-interior point of $\tilde{\mathcal{L}}$ [*Schaefer* 1974, p.96] with $TQ = Q$. Therefore the set $\mathcal{S} := \{T^k : k \in \mathbb{N}\}$ is an order-contractive semigroup of operators defined on $\tilde{\mathcal{L}}$. Due to *Nagel*'s theorem, $\mathcal{S}$ is mean ergodic, which entails $\|.\|$-convergence of μ_n to as T-invariant measure $\overline{\mu}$ as $n \to \infty$ (see, e.g. [*Schaefer* 1974, pp.343ff.]). For any $f \in \mathcal{M}$, the functional f' defined by $f'(\mu) := f(\overline{\mu})$ from $\mathcal{L}'$ to $\mathbb{R}$ is linear and continuous, since it satisfies $|f'(\mu)| \leq \|f\|\|\overline{\mu}\| \leq \|f\|\|\mu\|$. Thus we can extend f' to a linear, continuous functional $\overline{f} : \mathcal{L} \to \mathbb{R}$, which also fulfills $T^*\overline{f} = \overline{f}$. This follows by continuity from

$$T^*\overline{f}(\mu) = \overline{f}(T\mu) = f(\overline{T\mu}) = f(\overline{\mu}) = \overline{f}(\mu)$$

holding for all $\mu \in \mathcal{L}'$, because of the evident equation

$$\overline{T\mu} = \lim_{n \to \infty} \frac{1}{n} \sum_{k=1}^{n} T^k \mu = \lim_{n \to \infty} \mu_n = \overline{\mu} \quad \text{for all } \mu \in \mathcal{L}'.$$

Similarly, the $\sigma(\mathcal{M}, \mathcal{L})$-convergence of the sequence $\lim_{n \to \infty} \frac{1}{n} \sum_{k=0}^{n-1} T^{*k} f$ to the element $\overline{f}$ follows. *Luschgy* [1988] used this technique to show that for any semigroup $\{T_\theta : \theta \in G\}$ of positive contractions fixing $\mathcal{P}$, the sub-algebra

$$\mathcal{A}_G := \{f \in \mathcal{M} : T_\theta^* f = f \text{ for all } \theta \in G\}$$

of all G-invariant elements is sufficient. It is clear that all results of this section (and also some of the remainder of this chapter) remain true if one replaces the group G of transitions with a semigroup of positive contractions fixing $\mathcal{P}$. However, as it will be explained in the remark after the following theorem, the separate, different proof of Lemma 5.2 has its own right. $\qquad\qquad \triangle$

5.2. Sufficiency and invariance; full models

Now we shall show that the G-invariant generalized elements form a sufficient sub-algebra in $\mathcal{M}$, and present a criterion for the case that this sub-algebra coincides with the minimal sufficient sub-algebra. This result will be applied to the so-called full model, a situation where, roughly spoken, the experiment consists of all G-invariant probability measures. But at first we shall treat the case of a single transition:

Theorem 5.3 and Definition:

For $\theta \in G$ denote by

$$\mathcal{A}_\theta := \{ f \in \mathcal{M} : T_\theta^* f = f \} .$$

Then $\mathcal{A}_\theta$ is a sufficient sub-algebra of $\mathcal{M}$, the corresponding sufficient projection being defined by

$$\Pi_\theta : f \mapsto \lim_{n \to \infty} \frac{1}{n} \sum_{k=0}^{n-1} T_\theta^{*k} f$$

$$\mathcal{M} \to \mathcal{A}_\theta$$

Proof: Because of Lemma 5.1, $\mathcal{A}_\theta$ is a $\sigma(\mathcal{M}, \mathcal{L})$-closed sub-algebra of $\mathcal{M}$ containing e. By Lemma 5.2, the map Π_θ is well defined, obviously linear and positive, and fulfills $\Pi_\theta g = g$ for all $g \in \mathcal{A}_\theta$ as well as

$$(\Pi_\theta f)(P) = \lim_{n \to \infty} \frac{1}{n} \sum_{k=0}^{n-1} f(T_\theta^k P) = f(P) \quad \text{for all } f \in \mathcal{M} \text{ and all } P \in \mathcal{P},$$

whence the assertion follows with the help of Theorem 3.3. $\square$

Remark: Assume that $\theta : \Omega \to \Omega$ is model preserving (cf. remark preceding Lemma 5.1). Then for all $\varphi \in X$ there is a function $\overline{\varphi} \in X$ fulfilling

$$\frac{1}{n} \sum_{k=0}^{n-1} (\varphi \circ \theta^k)(\omega) \to \overline{\varphi}(\omega) \quad \mu\text{-almost surely for all } \mu \in \mathcal{L}$$

as $n \to \infty$ and $\overline{\varphi} \circ \theta = \overline{\varphi}$ (replace $Z, \hat{\mathcal{M}}, \hat{f}, r, \hat{T}, \hat{P}, \hat{\mu}$ with $\Omega, \mathcal{F}, \varphi, \overline{\varphi}, \theta, P, \mu$, respectively, in the proof of Lemma 5.2). Thus evidently $\Pi_\theta \dot{\varphi} = \dot{\overline{\varphi}}$, so that in this case we have $\Pi_\theta(\dot{X}) \subseteq \dot{X}$ (cf. Proposition 5.14 below!) $\triangle$

In the sequel we shall prove that every generalized estimator which is admissible w.r.t. a strictly convex loss function, has to be invariant under the group G of transitions. Furthermore, we shall specify a criterion for the property that, conversely, every generalized estimator invariant under G already is optimal (note that in this case invariance, sufficiency, and optimality are equivalent to each other, which in particular implies that every strictly convex loss function is universal):

Theorem 5.4 and Definition:

Denote by

$$\mathcal{A}_G := \bigcap_{\theta \in G} \mathcal{A}_\theta = \{f \in \mathcal{M} : T_\theta^* f = f \text{ for all } \theta \in G\}$$

the $\sigma(\mathcal{M}, \mathcal{L})$-closed sub-algebra of all invariant generalized estimators and by

$$\mathcal{V}_G := \{\sum_{i=1}^{n} f_i - T_{\theta_i}^* f_i : f_i \in \mathcal{M}, \theta_i \in G, 1 \leq i \leq n; n \in \mathbb{N}\} .$$

Then

(a) $\mathcal{A}_G$ is sufficient, and hence $\mathcal{A}_m \subseteq \mathcal{A}_G$;

(b) $\mathcal{A}_0 = \mathcal{A}_m = \mathcal{A}_G$ holds if and only if $\mathcal{V}_G$ is $\sigma(\mathcal{M}, \mathcal{L})$-dense in $\mathcal{V}$.

By Π_G we denote the sufficient projection onto $\mathcal{A}_G$.

Proof: (a) follows from Corollary 3.7 and Theorem 5.3.

(b): assume that $\mathcal{V}_G$ is $\sigma(\mathcal{M}, \mathcal{L})$-dense in $\mathcal{V}$. For $w \in \mathcal{V}_G$ and $f \in \mathcal{A}_G$ we have

$$(f \cdot w)(P) = \sum_{i=1}^{n}(f \cdot [f_i - T_{\theta_i}^* f_i])(P)$$

$$= \sum_{i=1}^{n}((f \cdot f_i)(P) - [T_{\theta_i}^*(f \cdot f_i)](P)) = 0 \quad \text{for all } P \in \mathcal{P},$$

whence $f \cdot w \in \mathcal{V}$ follows. Because of Proposition 1.5(b) the fact that $\mathcal{V}_G$ is $\sigma(\mathcal{M}, \mathcal{L})$-dense in $\mathcal{V}$ yields $\mathcal{A}_G \subseteq \mathcal{A}_0$. Corollary 4.6 now with the help of assertion (a) entails

$$\mathcal{A}_0 = \mathcal{A}_m = \mathcal{A}_G.$$

Conversely, assume that $\mathcal{A}_0$ coincides with $\mathcal{A}_G$. Then $\Pi_G v \in \mathcal{A}_0 \cap \mathcal{V} = \{o\}$ implies that $v = v - \Pi_G v$ belongs to the $\sigma(\mathcal{M}, \mathcal{L})$-closure $\overline{\mathcal{V}_G}$ of $\mathcal{V}_G$: indeed, let us first prove that

$$v - \Pi_\Theta v \in \overline{\mathcal{V}_G} \quad \text{holds for all } \Theta \in \Xi(G),$$

Π_Θ denoting the sufficient projection onto $\mathcal{A}_\Theta := \bigcap_{\theta \in \Theta} \mathcal{A}_\theta$. This assertion is shown by induction over the number of elements of $\Theta \in \Xi(G)$. For $\Theta = \{\theta\} \subseteq G$ we have

$$v - \Pi_\Theta v = v - \Pi_\theta v = \lim_{n \to \infty} \frac{1}{n} \sum_{k=0}^{n-1} (v - T_{\theta^k}^* v) \in \overline{\mathcal{V}_G}.$$

Assume now the claim being true for $\Theta \in \Xi(G)$; then it also holds for $\Theta' := \Theta \cup \{\theta\} \in \Xi(G)$, $\theta \in G$ being arbitrary, since by Theorem 3.5(b) we know

$$v - \Pi_{\Theta'} v = \lim_{n \to \infty} \frac{1}{n} \sum_{k=0}^{n-1} [v - (\Pi_\theta \circ \Pi_\Theta)^k v]$$

$$= \lim_{n \to \infty} \sum_{k=0}^{n-1} \sum_{j=0}^{k-1} (v_j - \Pi_\theta \circ \Pi_\Theta v_j),$$

where we have put $v_0 := \frac{1}{n} v$ and $v_j := \Pi_\theta(\Pi_\Theta v_{j-1})$ for $1 \le j \le k$. Now observe that for all $f \in \mathcal{M}$ we have $f - \Pi_\theta \circ \Pi_\Theta f \in \overline{\mathcal{V}_G}$: indeed, for $g = \Pi_\Theta f$ we infer via inductional assumption and the beginning of induction

$$f - \Pi_\theta \circ \Pi_\Theta f = f - \Pi_\Theta f + g - \Pi_\theta g \in \overline{\mathcal{V}_G}.$$

The claimed assertion finally follows from Theorem 3.6. $\qquad\qquad\qquad\qquad \square$

Remark: As we have noted already at the end of the preceding section, Theorem 5.4(a) is shown by *Luschgy* [1988] in a different way for a more general situation, where $\{T_\theta : \theta \in G\}$ is a semigroup of positive contractions on $\mathcal{L}$ which fix $\mathcal{P}$. In this setting *Luschgy* also shows that the converse holds: whenever $\mathcal{A}$ is a sufficient sub-algebra in $\mathcal{M}$, then there is a semigroup G of the above type such that $\mathcal{A} = \mathcal{A}_G$. To this end define

$$\{T_\theta : \theta \in G\} := \{T : \mathcal{L} \to \mathcal{L} : T \text{ is a positive contraction fixing } \mathcal{P}$$
$$\text{with } T^* f = f \text{ for all } f \in \mathcal{A}\}.$$

By definition of G, we have $\mathcal{A} \subseteq \{f \in \mathcal{M} : T_\theta^* f = f \text{ for all } \theta \in G\} = \mathcal{A}_G$. On the other hand, let $\Lambda : \mathcal{L} \to \mathcal{L}$ be such that $\Pi = \Lambda^*$ is the sufficient projection onto $\mathcal{A}$. Due to Theorem 3.3 and Proposition 2.5, the map Λ is a positive contraction fixing $\mathcal{P}$ and hence belongs to G because of $\Lambda^* g = \Pi g = g$ for all $g \in \mathcal{A}$. Thus for any $f \in \mathcal{A}_G$ we have $f = \Lambda^* f = \Pi f \in \mathcal{A}$. Therefore we have $\mathcal{A} = \mathcal{A}_G$. Let us note that in general the map Λ is no transition (cf. the following Proposition 5.5); hence the above argument seems to fail if G is a group of transitions. $\qquad\qquad \triangle$

The following fact is a consequence of Theorem 3.3. If G is generated by one element, it is obvious in the context of the remark after Lemma 5.2.

Proposition 5.5:
Let $\Lambda_G : \mathcal{L} \to \mathcal{L}$ be such that $\Pi_G = \Lambda_G^*$. Then $\Lambda_G \mu$ is G-invariant for all $\mu \in \mathcal{L}$.

Proof: By Theorem 3.3, we have $\Lambda_G \mu \in \mathcal{L}(\mathcal{A}_G)$ if $\mu \in \mathcal{L}'$. To be more precise, there is an element $g \in \mathcal{A}_G$, as well as a measure $Q = \sum_{i=1}^{n} \alpha_i P_i$ with $\alpha_i \geq 0$ and $P_i \in \mathcal{P}$ for $1 \leq i \leq n$, such that $\Lambda_G \mu = g \cdot Q$. By conseqence, for all $f \in \mathcal{M}$ and all $\theta \in G$ we obtain

$$f(T_\theta \Lambda_G \mu) = (T_\theta^* f)(g \cdot Q) = [(T_\theta^* f) \cdot g](Q) = [(T_\theta^* f) \cdot (T_\theta^* g)](Q)$$
$$= [T_\theta^*(f \cdot g)](Q) = (f \cdot g)(Q) = f(\Lambda_G \mu),$$

whence we deduce $\Lambda_G \mu = T_\theta(\Lambda_G \mu)$ for all $\theta \in G$ and all $\mu \in \mathcal{L}'$. A continuity argument now proves the assertion. $\qquad\square$

Let us turn now towards a frequently discussed special case: if $\tilde{T}_\theta : ca(\Omega, \mathcal{F}) \to ca(\Omega, \mathcal{F})$, $\theta \in G$, are transitions defined on $ca(\Omega, \mathcal{F})$, if

$$\mathcal{P} := \{P \in ca(\Omega, \mathcal{F})_+ : P(\Omega) = 1, \tilde{T}_\theta P = P \text{ for all } \theta \in G\},$$

and if

$$T_\theta := \tilde{T}_\theta|_{\mathcal{L}} \quad \text{for all } \theta \in G,$$

then we call the pair $(\mathcal{P}, G)$ "full model". In this case, invariance and optimality are equivalent, according to the following theorem. *Rüschendorf* [1987] derived an analoguous result in the classical setup using additional assumptions on G.

Theorem 5.6:

For a full model $(\mathcal{P}, G)$ we have $\mathcal{A}_0 = \mathcal{A}_m = \mathcal{A}_G$.

Proof: Let $v \in \mathcal{V}$ and $\mu \in \mathcal{L}_+$ with $\Lambda_G \mu \neq o$. Define

$$P := \overline{\Lambda_G \mu} = \frac{1}{\|\Lambda_G \mu\|} \Lambda_G \mu.$$

Proposition 5.5 yields invariance of the probability measure $P \in ca(\Omega, \mathcal{F})$. By fullness, we thus have $P \in \mathcal{P}$, whence we infer

$$(\Pi_G v)(\mu) = v(\Lambda_G \mu) = \|\Lambda_G \mu\| v(P) = 0.$$

The latter equations trivially also hold if $\Lambda_G \mu = o$. Therefore in any case we arrive at $\Pi_G v = o$. Hence $\mathcal{A}_G \cap \mathcal{V} = \{o\}$, implying the assertion via the remark after Theorem 4.7 and with the help of Theorem 5.4. $\qquad\square$

84

5.3. Calculation of Π_G, I.: special group properties

By the results specified above we see that in some cases it is possible to obtain optimal generalized estimators by application of the sufficient projection Π_G onto $\mathcal{A}_G$. Since the formula for Π_G derived from Theorems 3.5, 3.6, Corollary 3.7, and Theorem 5.3 may be rather inconvenient, we in the sequel calculate Π_G for finite, for commutative, and, finally, for amenable groups G:

Proposition 5.7:

Assume that G consists of exactly n elements, $n \in \mathbb{N}$. Then

(a) the sufficient projection Π_G onto $\mathcal{A}_G$ fulfills

$$\Pi_G = \frac{1}{n} \sum_{\theta \in G} T_\theta^*;$$

(b) moreover, we have $\mathcal{V}_G = \mathcal{V}$ if and only if $\mathcal{A}_0 = \mathcal{A}_m = \mathcal{A}_G$.

Proof: (a): because of

$$T_\tau^* \circ [\frac{1}{n} \sum_{\theta \in G} T_\theta^*] = \frac{1}{n} \sum_{\theta \in G} T_{\theta\tau}^* = \frac{1}{n} \sum_{\theta \in G} T_\theta^*,$$

of Lemma 5.1, and of $T_\theta^* f(P) = f(P)$ for all $(f, P) \in \mathcal{M} \times \mathcal{P}$, the equality

$$\frac{1}{n} \sum_{\theta \in G} T_\theta^* = \Pi_G$$

follows from the second part of the assertion in Proposition 3.4(b).
(b): if $\mathcal{A}_0 = \mathcal{A}_G$, then the relation $\Pi_G v = o$ for all $v \in \mathcal{V}$ entails

$$v = v - \Pi_G v = \sum_{\theta \in G} [(\frac{1}{n} v) - T_\theta^*(\frac{1}{n} v)] \in \mathcal{V}_G.$$

Theorem 5.4(b) yields the reverse implication. $\qquad\square$

Theorem 5.8:

If G is commutative then

$$(\Pi_{\theta_m} \circ \ldots \circ \Pi_{\theta_1}) f \to \Pi_G f \quad \text{w.r.t. } \sigma(\mathcal{M}, \mathcal{L}) \text{ as } \{\theta_1, \ldots, \theta_m\} \in \Xi(G),$$

where we view $\Xi(G)$ as a net that is directed with respect to set inclusion $\subseteq$.

Proof: Since $T_\theta^* \circ T_\tau^* = T_{\tau\theta}^* = T_\tau^* \circ T_\theta^*$ holds, we by Theorem 5.3 deduce $\Pi_\theta \circ T_\tau^* = T_\tau^* \circ \Pi_\theta$ and, again with the help of Theorem 5.3,

$$\Pi_\theta \circ \Pi_\tau = \Pi_\tau \circ \Pi_\theta \quad \text{for all } \tau \in G \text{ and all } \theta \in G.$$

Now Theorem 3.5(c) entails the assertion claimed. $\qquad\square$

A way to simplify the calculation of Π_G is to generalize Proposition 5.7 for infinite groups. To this end we shall employ the following amenability properties of G. Quite often in the remainder of this chapter we tacitly assume that G satisfies these conditions.

- Let $\mathcal{G}$ be a σ-field of sub-sets of G and η be a – not necessarily $(\sigma\text{-})$finite – measure defined on $\mathcal{G}$.

- The mapping

$$(\theta, \tau) \mapsto \theta\tau$$
$$G \times G \to G$$

  is $\mathcal{G} \otimes \mathcal{G}\text{-}\mathcal{G}$–measurable and η is right-invariant:

$$\eta(K\theta) = \eta(K) \quad \text{for all } K \in \mathcal{G} \text{ and all } \theta \in G.$$

- There exists a summing net $(K_i)_{i \in I}$ in $\mathcal{G}$; this means $\eta(K_i) < +\infty$ for all $i \in I$ and

$$\frac{1}{\eta(K_i)} \int_{K_i} [\phi(\theta) - \phi(\theta\tau)]\eta(d\theta) \to 0 \quad \text{as } i \in I$$

  for all bounded $\mathcal{G}$-measurable functions $\phi : G \to \mathbb{R}$, and all $\tau \in G$.

- G acts measurably on $\mathcal{E}$, which means that the map

$$\theta \mapsto f(T_\theta \mu)$$
$$G \to \mathbb{R}$$

  is $\mathcal{G}$-measurable for all $(f, \mu) \in \mathcal{M} \times \mathcal{L}$.

We shall deal with the first three of the conditions specified above in the remark following Theorem 5.12. Beforehand let us mention a sufficient condition that – in the situation described by the remark preceding Lemma 5.1 – G acts measurably on $\mathcal{E}$:

Proposition 5.9:

Assume that Ω is a σ-compact topological space and that $G \subseteq \Omega^\Omega$ is a topological group acting continuously on Ω which means that the maps

$$\theta \mapsto \theta(\omega)$$

$$G \to \Omega$$

are continuous for all $\omega \in \Omega$. Denote by $\mathcal{F}$, and $\mathcal{G}$, the system of *Borel* sets in Ω, and G, respectively. Furthermore assume that there is a countable, increasing sequence of sub-sets in Ω, say $\Omega_n \subseteq \Omega_{n+1}$, $n \in \mathbb{N}$, which are compact and metrizable with respect to the inherited topology, as well as G-invariant for all $n \in \mathbb{N}$. Then for all $\mu \in \mathcal{L}$ the map

$$\theta \mapsto T_\theta \mu$$

$$G \to \mathcal{L}$$

is $\mathcal{G} - \|.\|$-*Borel*–measurable. In particular G acts measurably on $\mathcal{E}$.

Proof: Let us denote the *Borel*-σ-field of the topological space Ω_n by $\mathcal{F}_n$, all $n \in \mathbb{N}$. For $\nu \in ca(\Omega, \mathcal{F})$ let $\nu_n := \nu|_{\mathcal{F}_n} \in ca(\Omega_n, \mathcal{F}_n)$. Next we show that

$$\theta \mapsto (T_\theta \mu)_n$$

$$G \to ca(\Omega_n, \mathcal{F}_n)$$

is continuous with respect to the $\sigma\big([C(\Omega_n)]^*, C(\Omega_n)\big)$-topology on $ca(\Omega_n, \mathcal{F}_n) = [C(\Omega_n)]^*$. Indeed, we have for $n \in \mathbb{N}$ and $f \in C(\Omega_n)$

$$\int f \, d(T_{\theta_k} \mu)_n = \int_{\Omega_n} f \, d(T_{\theta_k} \mu) = \int_{\Omega_n} f\big(\theta_k(\omega)\big) \, \mu(d\omega)$$

$$\to \int_{\Omega_n} f\big(\theta(\omega)\big) \, \mu(d\omega) = \int f \, d(T_\theta \mu)_n \quad \text{as } k \to \infty$$

provided that $\theta_k \to \theta$ as $k \to \infty$. This follows from the theorem of dominated convergence and from the observation that

$$\left| f\big(\theta_k(\omega)\big) \right| \le \sup_{\overline{\omega} \in \Omega_n} \left| f(\overline{\omega}) \right| < +\infty \quad \text{for all } \omega \in \Omega_n \text{ and all } k \in \mathbb{N}.$$

Since Ω_n is metrizable, $C(\Omega_n)$ is separable. Hence $\sigma\big([C(\Omega_n)]^*, C(\Omega_n)\big)$-continuity yields $\mathcal{G}-\|.\|$-*Borel*–measurability of the map $\theta \mapsto (T_\theta \mu)_n$. Now the equality

$$\|\mu\| = \sup_{n \in \mathbb{N}} |\mu|(\Omega_n) = \sup_{n \in \mathbb{N}} \|\mu_n\|_n \quad \text{for all } \mu \in ca(\Omega, \mathcal{F}),$$

$\|.\|_n$ being the variational norm on $ca(\Omega_n, \mathcal{F}_n)$, entails

$$\{\theta \in G : \|T_\theta \mu - \nu\| \le \alpha\} = \bigcap_{n \in \mathbb{N}} \{\theta \in G : \|(T_\theta \mu)_n - \nu_n\|_n < \alpha\} \in \mathcal{G},$$

whence the assertion follows. $\qquad\square$

Remark: The assumptions in Proposition 5.9 are fulfilled, e.g. if Ω is the dual space of a separable *Banach* space (equipped with the norm $\|.\|_\Omega$) and G is a group of isometries defined on Ω: in this case consider the weak*-topology on Ω, the norm topology on G, and define

$$\Omega_n := \{\omega \in \Omega : \|\omega\|_\Omega \leq n\} \quad \text{for all } n \in \mathbb{N}$$

(see, e.g. [*Dunford/Schwartz 1964*, pp.424ff.]). $\triangle$

The following result makes clear why we may view the operator Π_G as an averaging procedure over G:

Theorem 5.10:

Denote the sufficient projection onto $\mathcal{A}_G$ by Π_G. Then under the assumptions specified above

$$(\Pi_G f)(\mu) = \lim_{i \in I} \frac{1}{\eta(K_i)} \int_{K_i} f(T_\theta \mu)\, \eta(d\theta) \quad \text{holds for all } (f, \mu) \in \mathcal{M} \times \mathcal{L}.$$

Proof: The maps

$$W_i : (f, \mu) \mapsto \frac{1}{\eta(K_i)} \int_{K_i} f(T_\theta \mu)\, \eta(d\theta)$$
$$\mathcal{M} \times \mathcal{L} \to \mathbb{R}$$

all are well defined, bilinear, and bounded because of

$$|W_i(f, \mu)| \leq \|f\|\|\mu\| \quad \text{for all } (f, \mu) \in \mathcal{M} \times \mathcal{L}.$$

According to Theorem 1.7 there is an accumulation point $W : \mathcal{M} \times \mathcal{L} \to \mathbb{R}$ of the net $(W_i)_{i \in I}$ with respect to the topology of pointwise convergence in $\mathbb{R}^{\mathcal{M} \times \mathcal{L}}$. Via the definition

$$(\Pi f)(\mu) := W(f, \mu), \quad (f, \mu) \in \mathcal{M} \times \mathcal{L},$$

we obtain a linear, positive map $\Pi : \mathcal{M} \to \mathcal{A}_G$: indeed, for any sub-net I' of I we have

$$\frac{1}{\eta(K_i)} \int_{K_i} [f(T_\theta \mu) - f(T_\theta \circ T_\tau \mu)]\, \eta(d\theta) \;\to\; 0 \quad \text{as } i \in I'.$$

Furthermore, for all $g \in \mathcal{A}_G$ we obviously have $\Pi g = g$ as well as

$$(\Pi g)(P) = \lim_{i \in I'} \frac{1}{\eta(K_i)} \int_{K_i} g(T_\theta P)\, \eta(d\theta) = g(P) \quad \text{for all } (g, P) \in \mathcal{M} \times \mathcal{L}.$$

88

This yields $\Pi_G = \Pi$ by Theorem 3.3. In particular we know that there is only one accumulation point of $(W_i)_{i \in I}$, whence

$$W_i(f, \mu) \to (\Pi_G f)(\mu) \quad \text{as } i \in I$$

for all $(f, \mu) \in \mathcal{M} \times \mathcal{L}$ follows. $\qquad\square$

In order to prove an analogue of Theorem 5.4(b), which refers to the averaging procedure over G treated in the preceding theorem, we need an auxiliary result which even standing alone seems to be of some interest, since it exhibits the fact that this averaging procedure and multiplication with G-invariant generalized estimators are interchangeable:

Lemma 5.11:
Assume that $(f, g, P) \in \mathcal{M} \times \mathcal{A}_G \times \mathcal{P}$ and that $K \in \mathcal{G}$ fulfills $\eta(K) < +\infty$. Then

$$[g \cdot \int_K T_\theta^* f(.) \, \eta(d\theta)](P) = (f \cdot g)(P)\eta(K).$$

Proof: By inspecting the proof of Proposition 5.5 we see that $\mu := g \cdot P \in \mathcal{L}(\mathcal{A}_G)$ is G-invariant. Hence Corollary 1.6 yields

$$[g \cdot \int_K T_\theta^* f(.)] \, \eta(d\theta)](P) = \int_K f(T_\theta(g \cdot P)) \, \eta(d\theta) = f(g \cdot P)\eta(K). \qquad\square$$

In the following result we specify a variant of Theorem 5.4(b) involving integrals instead of sums.

Theorem 5.12:
Under the assumptions specified in Theorem 5.10, the set

$$\mathcal{V}_G' := \{\mu \mapsto \int_K f(\mu - T_\theta\mu) \, \eta(d\theta) : K \in \mathcal{G}, \, \eta(K) < +\infty, \, f \in \mathcal{M}\}$$

is $\sigma(\mathcal{M}, \mathcal{L})$-dense in $\mathcal{V}$ if and only if $\mathcal{A}_0 = \mathcal{A}_m = \mathcal{A}_G$.

Proof: Since

$$\int_K f(P - T_\theta P)\,\eta(d\theta) = 0 \quad \text{for all } (f, P) \in \mathcal{M} \times \mathcal{P},$$

we have $\mathcal{V}'_G \subseteq \mathcal{V}$. Furthermore, for $(g, w) \in \mathcal{A}_G \times \mathcal{V}'_G$ we by Lemma 5.11 obtain

$$(g \cdot w)(P) = (g \cdot f)(P)\eta(K) - (g \cdot f)(P) = 0 \quad \text{for all } P \in \mathcal{P}.$$

Therefore the fact that $\mathcal{V}'_G$ is $\sigma(\mathcal{M}, \mathcal{L})$-dense in $\mathcal{V}$ entails $\mathcal{A}_0 = \mathcal{A}_G$ with the help of Proposition 1.5(b).

Conversely, $\mathcal{A}_G = \mathcal{A}_0$ implies, by $\Pi_G v \in \mathcal{V} \cap \mathcal{A}_G = \mathcal{V} \cap \mathcal{A}_0 = \{o\}$ and Theorem 5.10

$$v(\mu) = v(\mu) - (\Pi_G v)(\mu) = \lim_{i \in I} \int_{K_i} (v_i - T_\theta^* v_i)(\mu)\,\eta(d\theta)$$

for all $\mu \in \mathcal{L}$, where we have put $v_i := \frac{1}{\eta(K_i)} v \in \mathcal{V}$, $i \in I$. $\qquad\square$

5.4. Calculation of Π_G, II.: invariance approximation property

We now turn towards the situation described in the remark preceding Lemma 5.1 again, where $G \subseteq \Omega^\Omega$: we call (Ω, G), or, to be more specific, $(\Omega, \mathcal{F}; G, \mathcal{G})$, a "dynamical system", if the map

$$(\omega, \theta) \mapsto \theta(\omega)$$

$$\Omega \times G \to \Omega$$

is $\mathcal{F} \otimes \mathcal{G}$–$\mathcal{F}$–measurable. Obviously G then consists of invertible maps from Ω onto itself that are bimeasurable with respect to $\mathcal{F}$. If $\mathcal{E}$ is a coherent experiment, then G acts measurably on $\mathcal{E}$, because

$$\dot\varphi(T_\theta\mu) = \int \varphi(\theta(\omega))\,\mu(d\omega)$$

holds due to *Fubini*'s theorem. To deal with invariance of generalized random variables we therefore need no assumption on a finer (e.g. topological) structure of G in order to show convergence of

$$\frac{1}{\eta(K_i)} \int_{K_i} f(T_\theta\mu)\,\eta(d\theta) \quad \text{as } i \in I.$$

By contrast, in the course of a classical treatment of invariance frequently amenable topological groups $G \subseteq \Omega^\Omega$ are considered (e.g. by [*Rüschendorf* 1987]), in which context $\mathcal{G}$

90

denotes the system of *Borel* sets in G. As usual, we shall say that such a dynamical system (Ω, G) satisfies the "*Emerson-Tempel'man* condition", if there is a constant $C > 0$ such that for the summing net $(K_i)_{i \in I}$ we have (cf. [*Tempel'man* 1972])

$$\eta(K_i^{-1} K_i) \leq C \eta(K_i) \quad \text{for all } i \in I.$$

If the topology on G fulfills the second countability axiom then $(K_i)_{i \in I}$ may be replaced by a summing sequence $(K_n)_{n \in \mathbb{N}}$ (see, e.g. [*Bondar/Milnes* 1981]).

In the sequel it will turn out that these assumptions are sufficient for the following condition to hold, which later on will prove important:

Definition:

Let G be a group consisting of model preserving transformations on $\mathcal{E}$ in the sense of the remark preceding Lemma 5.1. We shall say that $\mathcal{E}$ satisfies the "invariance-approximation condition with respect to G", if for all $f \in \mathcal{A}_G$ there exists a net of G-invariant functions $(\overline{\varphi}_i)_{i \in I}$ belonging to X such that

$$\dot{\overline{\varphi}}_i \to f \;\; \text{w.r.t. } \sigma(\mathcal{M}, \mathcal{L}) \text{ as } i \in I.$$

Proposition 5.13:

(a) Assume that the dynamical system (Ω, G) satisfies the conditions specified at the beginning of this section. Then

    (i) for all $\varphi \in X$ we have $\Pi_G \dot\varphi = \dot{\overline{\varphi}}$, where $\overline{\varphi} \in X$ is G-invariant and fulfills μ-almost surely

$$\frac{1}{\eta(K_n)} \int_{K_n} \varphi(\theta(\omega)) \, \eta(d\theta) \to \overline{\varphi}(\omega) \quad \text{as } n \to \infty.$$

    (ii) furthermore, we have for all $\mu \in \mathcal{L}$

$$\Lambda_G \mu(A) = \lim_{n \to \infty} \frac{1}{\eta(K_n)} \int_{K_n} \mu(\theta^{-1}(A)) \, \eta(d\theta) \quad \text{for all } A \in \mathcal{F}.$$

(b) Let G be a group consisting of model preserving transformations on $\mathcal{E}$. If for all $\varphi \in X$ there exists a G-invariant function $\overline{\varphi} \in X$ fulfilling

$$\Pi_G \dot\varphi = \dot{\overline{\varphi}},$$

then $\mathcal{E}$ satisfies the invariance-approximation condition with respect to G.

Proof: (a): According to Theorem 6.1 in [*Tempel'man 1972*], the sequence

$$\varphi_n(\omega) := \frac{1}{\eta(K_n)} \int_{K_n} \varphi\big(\theta(\omega)\big)\,\eta(d\theta) \quad \text{converges as } n \to \infty \text{ for all } \omega \in \Omega \setminus N,$$

where we have put

$$N := \{\omega \in \Omega : \liminf_{n \in \mathbb{N}} \varphi_n(\omega) < \limsup_{n \in \mathbb{N}} \varphi_n(\omega)\}.$$

N belongs to $\mathcal{N}$, because of $\nu(N) = 0$ for all G-invariant $\nu \in ca(\Omega, \mathcal{F})$. Since G is amenable, $\tau^{-1}(N) = N$ results for all $\tau \in G$. Thus by defining

$$\overline{\varphi}(\omega) := \begin{cases} \lim_{n \to \infty} \varphi_n(\omega), & \text{if } \omega \in \Omega \setminus N, \\ 0, & \text{if } \omega \in N, \end{cases}$$

we obtain a function $\overline{\varphi} \in X$ which is G-invariant. The observations that

$$|\varphi_n(\omega)| \leq |\varphi(\omega)| \quad \text{for all } n \in \mathbb{N} \text{ and all } \omega \in \Omega,$$

and that $\mu(N) = 0$, yield via $T_\theta^* \dot{\varphi} = (\varphi \circ \theta)^{\cdot}$

$$\Pi_G(\dot{\varphi})(\mu) = \lim_{n \to \infty} \frac{1}{\eta(K_n)} \int_{K_n} \big[\int_\Omega \varphi\big(\theta(\omega)\big)\,\mu(d\omega)\big]\,\eta(d\theta)$$

$$= \lim_{n \to \infty} \int_\Omega \varphi_n(\omega)\,\mu(d\omega) \;=\; \dot{\overline{\varphi}}(\mu),$$

thereby proving assertion (i) which in turn implies (ii) because of the equation

$$\Lambda_G \mu(A) = [\Pi_G(1_A^{\cdot})](\mu) = \overline{1_A^{\cdot}}(\mu)$$

$$= \lim_{n \to \infty} \frac{1}{\eta(K_n)} \int_\Omega \int_{K_n} 1_{\theta^{-1}(A)}(\omega)\,\eta(d\theta)\,\mu(d\omega).$$

To show (b) choose for $f \in \mathcal{A}_G$ a suitable net $(\varphi_i)_{i \in I}$ in X such that

$$\dot{\varphi}_i \to f \quad \text{w.r.t. } \sigma(\mathcal{M}, \mathcal{L}) \text{ as } i \in I.$$

Theorem 3.3 now entails

$$\dot{\overline{\varphi}}_i = \Pi_G \dot{\varphi}_i \to \Pi_G f = f \quad \text{w.r.t. } \sigma(\mathcal{M}, \mathcal{L}) \text{ as } i \in I. \qquad \square$$

Remark: If we assume that (Ω, G) is a dynamical system satisfying $\Pi_G(\dot{X}) \subseteq \dot{X}$, we can sharpen Proposition 5.13 considerably: if there is an arbitrary $(\sigma\text{-})$finite measure $\tilde{\eta}$ defined on $\mathcal{G}$ such that

$$\tilde{\eta}(K) = 0 \quad \text{if and only if} \quad \tilde{\eta}(K\theta) = 0 \text{ for all } \theta \in G,$$

(a variant of) assertion (a) in Proposition 5.13 readily follows. Indeed, since for $\Pi_G \dot{\varphi} = \dot{\psi}$, $\psi \in X$, we have $\dot{\psi} = \dot{\overline{\varphi}}$, where we define

$$\overline{\varphi}(\omega) := \begin{cases} \int_G \psi(\theta(\omega))\, \tilde{\eta}(d\theta), & \text{if } \omega \in \Omega \setminus N, \\ 0, & \text{if } \omega \in N, \end{cases}$$

as well as

$$
\begin{aligned}
N := \{ \omega \in \Omega : &\int_G \psi(\theta(\omega))\, \tilde{\eta}(d\theta) \neq \psi(\omega) \} \\
= \{ \omega \in \Omega : &\int_G | \int_G \psi(\theta(\omega))\, \tilde{\eta}(d\theta) - \psi(\tau(\omega))|\, \tilde{\eta}(d\tau) > 0 \} \in \mathcal{F}.
\end{aligned}
$$

Theorem 4 in [*Lehmann 1959*, p.225] now implies

$$\theta^{-1}(N) = N \quad \text{and} \quad \overline{\varphi} \circ \theta = \overline{\varphi} \quad \text{for all } \theta \in G$$

as well as $\mu(N) = 0$ for all $\mu \in ca(\Omega, \mathcal{F})$, thus a fortiori $\dot{\overline{\varphi}} = \dot{\psi}$. Note that in the preceding argument we without loss of generality already assumed $\tilde{\eta} \in ca(G, \mathcal{G})_+$. In the context of Proposition 5.13(a), the measure $\tilde{\eta} \in ca(G, \mathcal{G})_+$ defined by

$$\tilde{\eta}(K) := \sum_{n=1}^{\infty} \frac{\eta(K \cap K_n)}{\eta(K_n)} 2^{-n}, \quad K \in \mathcal{G},$$

fulfills the assumptions specified above. $\triangle$

5.5. Maximal invariants and metric transitivity

Definition:

Let G be a group of model preserving transformations on $\mathcal{E}$ and $\chi : \Omega \to \Omega$ be a $\mathcal{F}$-$\mathcal{F}$-measurable map which is G-invariant. χ is called "$\mathcal{P}$-maximal invariant with respect to G ", if

(i) χ furnishes $\mathcal{P}$ with a $\mathcal{F}$-$\sigma(\mathcal{L}, \mathcal{M})$-*Borel*–measurable parametrization, which means that we have $\mathcal{P} = \{P_{\chi(\omega)} : \omega \in \Omega\}$ as well as

$$\chi_f : \omega \mapsto f(P_{\chi(\omega)}) \in X$$
$$\Omega \to \mathbb{R}$$

for all $f \in \mathcal{M}$, and if

(ii) for all G-invariant $\overline{\varphi} \in X$ and all $\psi \in X$ we have

$$\int \overline{\varphi} \cdot \psi \, dP_{\chi(\omega)} = \overline{\varphi}(\omega) \int \psi \, dP_{\chi(\omega)} \quad \text{for all } \omega \in \Omega.$$

Remark: If χ is a maximal invariant with respect to G (i.e. for any G-invariant $\overline{\varphi} \in X$ there is a function $\varrho \in X$ fulfilling $\overline{\varphi} = \varrho \circ \chi$), which parametrizes $\mathcal{P}$ in an $\mathcal{F}$-$\sigma(\mathcal{L}, \mathcal{M})$-*Borel*–measurable way such that $P_{\chi(\omega)}(\{\omega' \in \Omega : \chi(\omega) = \chi(\omega')\}) = 1$ holds for all $\omega \in \Omega$, then χ is a $\mathcal{P}$-maximal invariant with respect to G: indeed, for G-invariant $\overline{\varphi} \in X$ and arbitrary $\psi \in X$ we have

$$\int_\Omega \overline{\varphi} \cdot \psi \, dP_{\chi(\omega)} = \int_{\{\omega' \in \Omega : \chi(\omega') = \chi(\omega)\}} \varrho(\chi(\omega'))\psi(\omega') \, P_{\chi(\omega)}(d\omega') = \overline{\varphi}(\omega) \int_\Omega \psi \, dP_{\chi(\omega)}. \quad \triangle$$

The existence of a $\mathcal{P}$-maximal invariant with respect to G entails that every $P \in \mathcal{P}$ is metrically transitive with respect to G: for G-invariant $\psi = \overline{\varphi} = 1_A$ with $A \in \mathcal{F}$, property (ii) of χ yields $P_{\chi(\omega)}(A) = 1_A(\omega)P_{\chi(\omega)}(A)$ for all $\omega \in \Omega$, thus $P(A) = [P(A)]^2$, whence we infer $P(A) \in \{0, 1\}$. Therefore every G-invariant estimator is P-almost surely constant and hence optimal in the classical sense. The theorem specified below shows this to be true even for generalized estimators. Moreover, it exhibits the fact that in this situation there necessarily exist some non-trivial optimal estimators. Interestingly enough, every G-invariant generalized random variable here is already an equivalence class of a G-invariant random variable in the classical sense:

94

Theorem 5.14:

Assume that $\mathcal{E}$ fulfills the invariance-approximation condition with respect to G. Let χ be a $\mathcal{P}$-maximal invariant with respect to G. Then

(a) $\Pi_G f = \dot{\chi}_f$ holds for all $f \in \mathcal{M}$;

(b) $\mathcal{A}_0 \cap \dot{X} = \mathcal{A}_0 = \mathcal{A}_m = \mathcal{A}_G$;

(c) $\mathbb{R} \cdot e \neq \mathcal{A}_0 \cap \dot{X} = \mathcal{A}_0$ holds provided $\mathcal{P}$ consists of more than one probability measure.

Proof: (a): at first let us show that $(f \cdot g)(P) = f(P)g(P)$ holds for all $(f, g) \in \mathcal{M} \times \mathcal{A}_G$. Property (ii) of χ in the definition given above entails that, for G-invariant $\overline{\varphi} \in X$ and arbitrary $\psi \in X$, we have

$$(\dot{\overline{\varphi}} \cdot \dot{\psi})(P) = \dot{\overline{\varphi}}(P)\dot{\psi}(P).$$

This follows from the observation that

$$\dot{\overline{\varphi}}(P_{\chi(\omega)}) = \int \overline{\varphi} \cdot 1_\Omega \, dP_{\chi(\omega)} = \overline{\varphi}(\omega) \cdot 1 \quad \text{for all } \omega \in \Omega.$$

Now note that for $f \in \mathcal{M}$ and for $g \in \mathcal{A}_G$ there exists, by assumption, a net $(\overline{\varphi}_i)_{i \in I}$ in X of G-invariant functions, as well as a net of functions $(\psi_j)_{j \in J}$ in X, such that

$$\dot{\overline{\varphi}}_i \to g \ \text{ w.r.t. } \ \sigma(\mathcal{M}, \mathcal{L}) \text{ as } i \in I \quad \text{and} \quad \psi_j \to f \ \text{ w.r.t. } \ \sigma(\mathcal{M}, \mathcal{L}) \text{ as } j \in J,$$

whence from Proposition 1.5(b) $(f \cdot \dot{\psi}_j)(P) = \lim_{i \in I}(\dot{\overline{\varphi}}_i \cdot \dot{\psi}_j)(P) = [\lim_{i \in I} \dot{\overline{\varphi}}_i(P)]\dot{\psi}_j(P) = f(P)\dot{\psi}_j(P)$ for all $j \in J$, and finally, again by Proposition 1.5(b),

$$(f \cdot g)(P) = \lim_{j \in J}(f \cdot \dot{\psi}_j)(P) = \lim_{j \in J}[f(P)\dot{\psi}_j(P)] = f(P)g(P)$$

follows. Furthermore, for all $f \in \mathcal{M}$ the fact that χ_f is G-invariant yields

$$\dot{\chi}_f(P_{\chi(\omega)}) = \chi_f(\omega) = f(P_{\chi(\omega)}) \quad \text{for all } \omega \in \Omega,$$

whence we conclude

$$(f \cdot g)(P) = f(P)g(P) = \dot{\chi}_f(P)g(P) = (\dot{\chi}_f \cdot g)(P). \tag{$*$}$$

The map

$$f \mapsto \dot{\chi}_f$$
$$\mathcal{M} \to \mathcal{A}_G$$

is clearly linear and positive (note that $T_\theta^*(\dot\chi_f) = \big(f(P_{\chi\circ\theta(.)})\big)^\cdot = \dot\chi_f$). Moreover we have $\dot\chi_e = e$ and, by $(*)$ and Proposition 3.1,

$$\dot\chi_{\dot\chi_f} = \dot\chi_f.$$

Therefore Proposition 3.4(b) entails $\Pi_G f = \dot\chi_f \in \mathcal{A}_G \cap \dot{X}$.

(b): let $v \in \mathcal{V}$. Because of

$$\dot\chi_v(\mu) = \int v(P_{\chi(\omega)})\,\mu(d\omega) = 0 \quad \text{for all } \mu \in \mathcal{L},$$

we infer $\Pi_G v = o$, which means $\mathcal{A}_G \cap \mathcal{V} = \{0\}$ and thus $\mathcal{A}_G \subseteq \mathcal{A}_0$, whence we by (a) obtain $\mathcal{A}_0 = \mathcal{A}_G = \mathcal{A}_G \cap \dot{X}$.

(c): since $\{P, Q\} \subseteq \mathcal{P}$ is linearly independent provided $P \neq Q$ there exists a functional $f \in \mathcal{M}$ which is not constant on $\mathcal{P}$. Hence $\dot\chi_f \in (\mathcal{A}_0 \cap X) \setminus \mathbb{R} \cdot e$ follows. $\qquad\square$

Remark: Theorem 5.14 suggests the conjecture that experiments satisfying the invariance-approximation condition with respect to G, which admit a $\mathcal{P}$-maximal invariant, necessarily are coherent. This is indeed true for the situation described in the remark preceding Theorem 5.14, if one makes an additional measurability assumption (see Corollary A.3 in the appendix). The subsequent example however shows that in general this conjecture does not hold (see the example following Proposition 1.1; cf. [*Rogge 1972*]): $\qquad\triangle$

Example: Let $\Omega = \mathbb{C}$, $\mathcal{F}$ be the system of *Borel* sets in Ω, $G = \{\theta \in \Omega : |\theta| = 1\}$, G acting multiplicatively (and thus continuously) on Ω. For $r \geq 0$ and $A \in \mathcal{F}$ define

$$P_r(A) := \frac{1}{2\pi} \int_0^{2\pi} 1_A(re^{it})\,dt$$

and let $\mathcal{P} = \{P_r : r \geq 0\}$. G and $\mathcal{E}$ satisfy the assumptions specified in Proposition 5.13(a) and therefore those in Theorem 5.14 are fulfilled. The map

$$\chi : \omega \mapsto |\omega|$$
$$\Omega \to \Omega$$

is a maximal invariant with respect to G as well as

$$\omega \mapsto P_{|\omega|}$$
$$\Omega \to \mathcal{P}$$

is a $\mathcal{F}\text{-}\sigma(\mathcal{L}, \mathcal{M})$-*Borel*–measurable parametrization of $\mathcal{P}$: indeed, if we denote by

$$\Omega_n := \{\omega \in \Omega : |\omega| \leq n\} \text{ and by } \mathcal{F}_n \text{ the system of } Borel \text{ sets in } \Omega_n\,, \ n \in \mathbb{N},$$

then for arbitrary $\varphi \in C(\Omega_n)$

$$\varphi(r_k e^{it}) \to \varphi(r e^{it}) \quad \text{as } k \to \infty \text{ uniformly in } t \in [0, 2\pi],$$

provided that $r_k \to r$ as $k \to \infty$. Hence

$$\int_\Omega \varphi \, dP_{r_k} \to \int_\Omega \varphi \, dP_r \quad \text{as } k \to \infty,$$

yielding $\sigma\big([C(\Omega_n)]^*, C(\Omega_n)\big)$-*Borel*–measurability of the map

$$r \mapsto P_r|_{\mathcal{F}_n}$$
$$\mathbb{R}_+ \to [C(\Omega_n)]^*.$$

By analogy to the proof of Proposition 5.9,

$$\omega \mapsto P_{\chi(\omega)}$$
$$\Omega \to \mathcal{L}$$

therefore is $\mathcal{F}$–$\|.\|$-*Borel*–measurable and thus a fortiori is $\mathcal{F}$–$\sigma(\mathcal{L}, \mathcal{M})$-*Borel*–measurable. Hence Theorem 5.14 entails $\mathcal{A}_G = \mathcal{A}_0 = \mathcal{A}_m \neq \mathbb{R} \cdot e$. $\qquad \bowtie$

5.6. Banach space results

Let us close this chapter again by treating the higher dimensional case: let B be a *Banach* space. The $\sigma(\mathcal{M}, \mathcal{L})$-$\sigma(\mathcal{M}, \mathcal{L})$-continuous, linear map $T_\theta^* : \mathcal{M} \to \mathcal{M}$ induces, according to Proposition 1.9, a linear map $(T_\theta^*)^B : \mathcal{M}^B \to \mathcal{M}^B$, which in this case is of the form

$$((T_\theta^*)^B F)(\mu) = F(T_\theta \mu), \quad F \in \mathcal{M}^B, \, \mu \in \mathcal{L}, \, \theta \in G,$$

as one easily conceives. Therefore we have for all $\theta \in G$

$$\mathcal{A}_\theta^B := \{F \in \mathcal{M} : (T_\theta^*)^B F = F\} = (\mathcal{A}_\theta)^B$$

as well as

$$\mathcal{A}_G^B := \bigcap_{\theta \in G} \mathcal{A}_\theta^B = (\mathcal{A}_G)^B.$$

As an immediate consequence we obtain the following

Theorem 5.15:

(a) $\mathcal{A}_m{}^B$ is contained in $\mathcal{A}_G^B$.

(b) The assertions
    (i) $\mathcal{A}_0{}^B = \mathcal{A}_m{}^B = \mathcal{A}_G^B$;
    (ii) $[\mathcal{V}_G]^B$ is $\sigma(\mathcal{M},\mathcal{L})/\sigma(B,B^*)$-dense in $\mathcal{V}^B$;
    (iii) $\mathcal{V}_G$ is $\sigma(\mathcal{M},\mathcal{L})$-dense in $\mathcal{V}$;
    are equivalent.

(c) For the full model $\mathcal{E}$ we have $\mathcal{A}_0{}^B = \mathcal{A}_m{}^B = \mathcal{A}_G^B$.

Proof: The claimed assertions are easily derived from Theorems 5.4 and 5.6. $\qquad\square$

The calculation of the map $(\Pi_G)^B : \mathcal{M}^B \to \mathcal{A}_G^B$ – which exists due to Proposition 1.9 and Theorem 3.3 – can be extended from univariate to multivariate settings in a similarly evident way:

Theorem 5.16:

Let $(F,\mu) \in \mathcal{M}^B \times \mathcal{L}$. Then

(a) for $\theta \in G$ we have

$$\frac{1}{n}\sum_{k=0}^{n-1}[(T_{\theta^k}^*)^B F](\mu) \to [(\Pi_\theta)^B F](\mu) \ \ \text{w.r.t.}\ \ \sigma(B,B^*) \ \text{as}\ n \to \infty;$$

(b) if G consists of n elements then

$$(\Pi_G)^B F = \frac{1}{n}\sum_{\theta \in G}(T_\theta^*)^B F;$$

(c) if G is commutative then

$$[(\Pi_{\theta_n})^B \circ \ldots \circ (\Pi_{\theta_1})^B F](\mu) \to [(\Pi_G)^B F](\mu) \ \ \text{w.r.t.}\ \ \sigma(B,B^*)$$
$$\text{as}\ \{\theta_1,\ldots,\theta_n\} \in \Xi(G);$$

(d) if the assumptions specified in Theorem 5.10 are met and B is separable then

$$\frac{1}{\eta(K_i)}\int_{K_i}[(T_\theta^*)^B F](\mu)\,\eta(d\theta) \to [(\Pi_G)^B F](\mu) \ \ \text{w.r.t.}\ \ \sigma(B,B^*)\ \text{as}\ i \in I,$$

the integrals being understood in *Bochner's* sense.

98

Proof: The assertions claimed in (a),(b),(c) follow by Propositions 1.9, 5.7, and Theorems 5.3, 5.8, respectively. As far as (d) is concerned, observe that the map

$$w : \theta \mapsto [(T_\theta^*)^B F](\mu)$$
$$G \to B$$

is $\mathcal{G}$–$\sigma(B, B^*)$-*Borel*–measurable, since $\theta \mapsto \langle w(\theta), y \rangle = \langle F, y \rangle (T_\theta \mu)$ is $\mathcal{G}$-measurable for all $y \in B^*$ by assumption. Now separability of B implies $\mathcal{G}$–$\|.\|$-*Borel*–measurability, and thus *Bochner* integrability of the map w (note that $|||w(\theta)||| \leq |||F||| \, \|\mu\|$ holds for all $\theta \in G$). Therefore the expressions

$$x_i := \frac{1}{\eta(K_i)} \int_{K_i} w(\theta) \, \eta(d\theta) = \frac{1}{\eta(K_i)} \int_{K_i} [(T_\theta^*)^B F](\mu) \, \eta(d\theta) \in B$$

for all $i \in I$ make sense, and Theorem 5.10 together with Proposition 1.9 entails

$$\langle x_i, y \rangle = \frac{1}{\eta(K_i)} \int_{K_i} [T_\theta^* \langle F, y \rangle](\mu) \, \eta(d\theta) \to \langle [(\Pi_G)^B F](\mu), y \rangle \quad \text{as } i \in I$$

for all $y \in B^*$. $\qquad\qquad\square$

Finally also Theorem 5.14 carries over to multivariate situations:

Theorem 5.17:

If the assumptions specified in Theorem 5.14 are in force then

(a) $(\Pi_G)^B F = \dot{\chi}_F \in \dot{X}^B$, where $\chi_F(\omega) := F(P_{\chi(\omega)})$, all $(\omega, F) \in \Omega \times \mathcal{M}^B$;

(b) $\mathcal{A}_0{}^B = \mathcal{A}_m{}^B = \mathcal{A}_G^B$;

(c) $B \cdot e \neq \mathcal{A}_0{}^B \cap \dot{X}^B = \mathcal{A}_0{}^B$ provided $\mathcal{P}$ contains more than one probability measure.

Proof: Because of

$$\langle \chi_F(\omega), y \rangle = \chi_{\langle F, y \rangle}(\omega) \quad \text{for all } (\omega, F, y) \in \Omega \times \mathcal{M}^B \times B^*,$$

the assertions claimed follow by Proposition 1.9 and Theorem 5.14. $\qquad\qquad\square$

Remark: If G satisfies the *Emerson-Tempel'man* condition, then for all $\Phi \in X^B$ we have $(\Pi_G)^B \dot{\Phi} = \dot{\overline{\Phi}}$, where $\overline{\Phi} \in X^B$ is G-invariant and fulfills μ-almost surely

$$|||\frac{1}{\eta(K_n)} \int_{K_n} \Phi(\theta(\omega)) \, \eta(d\theta) - \overline{\Phi}(\omega)||| \to 0 \quad \text{as } n \to \infty$$

(see Theorem 6.2 in [*Tempel'man 1972*]). $\qquad\qquad\triangle$

Appendix: Alternative representation of M-spaces

To define $\mathcal{M}$, *Torgersen* [1980] chose an approach completely different from the considerations exhibited above. His concept enables us to generalize certain classes of unbounded random variables, e.g. functions which are p–fold integrable with respect to all $P \in \mathcal{P}$, where $1 \leq p \leq +\infty$. In the sequel we shall show how this notion carries over to the space $^B\mathcal{M}$ of generalized random elements and prove that both concepts coincide. To this end we need some further definitions: let $\mu \in \mathcal{L}_+$ and consider two maps $\Phi, \Phi' : \Omega \to B$ which are *Bochner* integrable with respect to μ. We write

$$\Phi \overset{\mu}{\sim} \Phi', \quad \text{if} \quad \Phi(\omega) = \Phi'(\omega) \quad \mu\text{-almost surely.}$$

If $1 \leq p < +\infty$, the symbol $L_B^p(\mu)$ denotes the set of all $\overset{\mu}{\sim}$-equivalence classes H_μ of maps $\Phi : \Omega \to B$, which are *Bochner* integrable with respect to μ and which fulfill

$$\int |||\Phi(\omega)|||^p \, \mu(d\omega) < +\infty .$$

$L_B^p(\mu)$ is a *Banach* space, if we equip it with the norm given by the expression

$$\|H_\mu\|_{p,\mu} := \Big[\int |||\Phi(\omega)|||^p \, \mu(d\omega) \Big]^{1/p} ,$$

which is the same for all representative elements $\Phi \in H_\mu$ of the class H_μ.

Considering now maps from Ω to B^*, put for any two elements Ψ and Ψ' of $M_{B*}^\infty(B) \supseteq {}^B X$ (cf. remark preceding Proposition 1.8 in section 1.3.)

$$\Psi \overset{\mu}{\sim}{}_* \Psi', \quad \text{if} \quad \langle x, \Psi(\omega) \rangle = \langle x, \Psi'(\omega) \rangle \quad \mu\text{-almost surely for all } x \in B .$$

For a $\overset{\mu}{\sim}_*$-equivalence class F_μ of $\Psi' \in M_{B*}^\infty(B)$, we define

$$\|F_\mu\|_{\infty,\mu} := \inf_{\Psi \in F_\mu} \big(\inf\{\alpha \geq 0 : \mu^\star(\{\omega \in \Omega : |||\Psi(\omega)||| > \alpha\}) = 0\} \big) ,$$

where $\mu^\star$ denotes the outer measure corresponding to μ; then the inequality $\|F_\mu\|_{\infty,\mu} \leq |||\Psi'|||_\infty \|\mu\| < +\infty$ holds. The set $L_{B*}^\infty(\mu, B)$ of all such equivalence classes F_μ, furnished by the norm $\|.\|_{\infty,\mu}$, is a *Banach* space which is isometrically isomorphic to the space $[L_B^1(\mu)]^*$. Here the duality is given by the expression

$$\langle H_\mu, F_\mu \rangle_\mu := \int \langle \Phi(\omega), \Psi(\omega) \rangle \, \mu(d\omega), \quad (H_\mu, F_\mu) \in L_B^1(\mu) \times L_{B*}^\infty(\mu, B),$$

where $\Phi \in H_\mu$ and $\Psi \in F_\mu$ are arbitrarily chosen representatives of their classes [*Ionescu-Tulcea/Ionescu-Tulcea* 1969, pp.15, 76ff., and 94f.].

As usual, the symbol

$$\bigotimes_{P \in \mathcal{P}} L_{B*}^\infty(P, B)$$

denotes the product of the *Banach* spaces $L_{B*}^\infty(P, B), P \in \mathcal{P}$. For $\tilde{F} = (F_P)_{P \in \mathcal{P}} \in \bigotimes_{P \in \mathcal{P}} L_{B*}^\infty(P, B)$ we define

$$\|\tilde{F}\|_{\mathcal{P}} := \sup_{P \in \mathcal{P}} \|F_P\|_{\infty, P} \leq +\infty.$$

In the case $B = \mathbb{R}$ we denote by $\leq$ the natural ordering in the product $\bigotimes_{P \in \mathcal{P}} L^\infty(P)$ of the *Banach* lattices $L^\infty(P)$, $P \in \mathcal{P}$, i.e.

$$(f_P)_{P \in \mathcal{P}} \leq (g_P)_{P \in \mathcal{P}} \quad \text{if and only if} \quad \varphi(\omega) \leq \psi(\omega) \ \ P\text{-almost surely}$$

for all $(\varphi, \psi) \in f_P \times g_P, P \in \mathcal{P}$.

Theorem A.1:

Let

$$^B\tilde{\mathcal{M}} := \{ \tilde{F} = (F_P)_{P \in \mathcal{P}} \in \bigotimes_{P \in \mathcal{P}} L_{B*}^\infty(P, B) : \bigcap_{P \in \mathcal{Q}} F_P \neq \emptyset \text{ for all } \mathcal{Q} \in \Xi(\mathcal{P})$$
$$\text{and } \|\tilde{F}\|_{\mathcal{P}} < +\infty \}.$$

Then $^B\mathcal{M}$ is isometrically isomorphic to $^B\tilde{\mathcal{M}}$, provided $^B\tilde{\mathcal{M}}$ is furnished by the norm $\|.\|_{\mathcal{P}}$. In the case $B = \mathbb{R}$, this isomorphism also preserves order.

Proof: (a) Let $\tilde{F} = (F_P)_{P \in \mathcal{P}} \in {}^B\tilde{\mathcal{M}}$ and $\mu \in \mathcal{L}'$ be such that $|\mu| \leq Q := \sum_{i=1}^n \alpha_i P_i$, $\alpha_i \geq 0$, $P_i \in \mathcal{P}$, $1 \leq i \leq n$. We put

$$F'(\mu) :- \int \Psi \, d\mu \, , \ \Psi \in \bigcap_{i=1}^n F_{P_i} \subseteq M_{B*}^\infty(B),$$

101

where

$$\int \Psi \, d\mu : x \mapsto \int \langle x, \Psi(\omega) \rangle \, \mu(d\omega)$$

$$B \to \mathbb{R}$$

denotes the $\sigma(B^*, B)$–*Pettis* integral as in the remark preceding Proposition 1.8. Then $F'(\mu)$ is a well defined linear functional on B: indeed, if also $|\mu| \le Q' := \sum_{j=1}^{m} \beta_j P'_j$ for some $\beta_j \ge 0$ and $P'_j \in \mathcal{P}$, $1 \le j \le m$, then evidently $|\mu| \le \overline{Q} := Q + Q'$. Now denote by $\mathcal{Q} := \{P_1, \ldots, P_n, P'_1, \ldots, P'_m\} \in \Xi(\mathcal{P})$. If we pick

$$(\Psi', \overline{\Psi}) \in \bigcap_{j=1}^{n} F_{P'_j} \times \bigcap_{P \in \mathcal{Q}} F_P$$

and if $\varphi \in X$, and $\varphi' \in X$, respectively, denote versions of the *Radon/Nikodym* densities $\frac{d\mu}{dQ}$, and $\frac{d\mu}{dQ'}$, respectively (which exist due to $|\mu| \le Q$ and $|\mu| \le Q'$), then for all $x \in B$ we have

$$\begin{aligned}
\int \langle x, \Psi \rangle \, d\mu &= \sum_{i=1}^{n} \alpha_i \int \langle x, \Psi \rangle \cdot \varphi \, dP_i \\
&= \sum_{i=1}^{n} \alpha_i \int \langle x, \overline{\Psi} \rangle \cdot \varphi \, dP_i \\
&= \int \langle x, \overline{\Psi} \rangle \cdot \varphi \, dQ = \int \langle x, \overline{\Psi} \rangle \, d\mu \\
&= \sum_{j=1}^{m} \beta_j \int \langle x, \overline{\Psi} \rangle \cdot \varphi' \, dP'_j = \int \langle x, \Psi' \rangle \, d\mu.
\end{aligned}$$

These equalities also make clear, that $\int \Psi d\mu = \int \Psi' d\mu$ holds whenever $\{\Psi, \Psi'\} \subseteq F_\mu$. Similarly we infer that for all $(\alpha, \nu) \in \mathbb{R} \times \mathcal{L}'$

$$F'(\alpha\mu + \nu) = \alpha F'(\mu) + F'(\nu).$$

Furthermore for all $x \in B$

$$|F'(\mu)(x)| = \left| \int \langle x, \Psi \rangle \, d\mu \right| = \left| \int \langle x, \Psi \rangle \cdot h \, dQ \right|$$

$$\le \int |\langle x, \Psi \rangle| \cdot |h| \, dQ = \sum_{i=1}^{n} \alpha_i \int |\langle x, F_{P_i} \rangle| \cdot |h| \, dP_i$$

$$\le |||x||| \, \|\tilde{F}\|_{\mathcal{P}} \int |h| \, dQ \le |||x||| \, \|\tilde{F}\|_{\mathcal{P}} \|\mu\|,$$

entailing

$$|||F'(\mu)||| \le \|\tilde{F}\|_{\mathcal{P}} \|\mu\|,$$

which means $F'(\mu) \in B^*$. According to Proposition 1.1(b) we may and do extend F' in a unique way to a linear map $F : \mathcal{L} \to B^*$ fulfilling $F \in {}^B\mathcal{M}$ or, to be more specific,

$$\||F|\| \leq \|\tilde{F}\|_\mathcal{P} < +\infty.$$

We thus constructed a map

$$\tilde{F} \mapsto F$$
$$^B\tilde{\mathcal{M}} \to {}^B\mathcal{M}$$

that is clearly linear.

(b) Now let $F \in {}^B\mathcal{M}$ and $\mu \in \mathcal{L}_+$. Then we claim that

there is exactly one element $F_\mu \in L^\infty_{B*}(\mu, B)$ fulfilling

$$\|F_\mu\|_{\infty,\mu} \leq \||F|\| \,\|\mu\| \quad \text{and} \quad (\dot{\varphi} \cdot F)(\mu) = \int \varphi \cdot \Psi \, d\mu \quad \text{for all } (\varphi, \Psi) \in X \times F_\mu. \qquad (*)$$

Indeed, choose a bounded net $(\Psi_i)_{i\in I}$ in $X^{B^*} \subseteq M^\infty_{B*}(B)$ approximating F, i.e. $\||\Psi_i|\|_\infty \leq \||F|\|$ for all $i \in I$ and $\Psi_i \to F$ w.r.t. $\sigma(\mathcal{M}, \mathcal{L})/\sigma(B^*, B)$ as $i \in I$ (see Proposition 1.8). Denote by $(F_{i,\mu})_{i\in I}$ the corresponding net of $\overset{\mu}{\sim}*$-equivalence classes in $L^\infty_{B*}(\mu, B)$. According to *Alaoglu*'s theorem,

$$\||F_{i,\mu}|\|_{\infty,\mu} \leq \||\Psi_i|\|_\infty \|\mu\| \leq \||F|\| \,\|\mu\| \quad \text{for all } i \in I$$

implies the existence of a sub-net $(F_{j,\mu})_{j\in J}$ and of an element $F_\mu \in L^\infty_{B*}(\mu, B)$ fulfilling $\|F_\mu\|_{\infty,\mu} \leq \||F|\| \,\|\mu\|$ such that

$$F_{j,\mu} \to F_\mu \quad \text{w.r.t. } \sigma\big(L^\infty_{B*}(\mu, B), L^1_B(\mu)\big) \text{ as } j \in J,$$

which in particular entails that

$$\langle x \cdot f_\mu, F_{j,\mu}\rangle_\mu \to \langle x \cdot f_\mu, F_\mu\rangle_\mu \quad \text{as } j \in J \text{ for all } f_\mu \in L^1(\mu),$$

since $x \cdot f_\mu \in L^1_B(\mu)$ holds for all $x \in B$. Now if $\varphi \in X$ and $f_\mu \in L^1(\mu)$ fulfill $\varphi \in f_\mu$, then we on one hand have

$$\langle x, (\dot{\varphi} \cdot \dot{\Psi}_j)(\mu)\rangle = \langle x, \dot{\Psi}_j(\dot{\varphi} \cdot \mu)\rangle = \int \langle x, \Psi_j\rangle \cdot \varphi \, d\mu$$
$$= \langle x \cdot f_\mu, F_{j,\mu}\rangle_\mu \to \langle x \cdot f_\mu, F_\mu\rangle_\mu$$
$$= \langle x, \int \varphi \cdot \Psi \, d\mu\rangle \quad \text{as } j \in J \text{ for all } \Psi \in F_\mu$$

and, on the other hand,

$$\langle x, \dot{\Psi}_j \rangle \to \langle x, F \rangle \quad \text{w.r.t.} \ \sigma(\mathcal{M}, \mathcal{L}) \ \text{as} \ j \in J$$

together with Proposition 1.5(b) yields

$$\langle x, (\dot{\varphi} \cdot \dot{\Psi}_j)(\mu) \rangle = [\dot{\varphi} \cdot \langle x, \Psi_j \rangle^{\cdot}](\mu) = [\dot{\varphi} \cdot \langle x, \dot{\Psi}_j \rangle](\mu)$$
$$\to [\dot{\varphi} \langle x, F \rangle](\mu) = \langle x, F(\dot{\varphi} \cdot \mu) \rangle = \langle x, (\dot{\varphi} \cdot F)(\mu) \rangle \quad \text{as} \ j \in J.$$

Uniqueness of F_μ is evident so that $(*)$ is proven. Given $F \in {}^B\mathcal{M}$, now choose $F_P \in L^\infty_{B*}(P, B)$ such that $(*)$ is satisfied for $\mu = P \in \mathcal{L}_+$, all $P \in \mathcal{P}$, and define

$$\tilde{F} := (F_P)_{P \in \mathcal{P}} \in \bigotimes_{P \in \mathcal{P}} L^\infty_{B*}(P, B).$$

Then the map $F \mapsto \tilde{F}$ is well defined and evidently linear due to assertion $(*)$; furthermore its values $\tilde{F}$ belong to ${}^B\tilde{\mathcal{M}}$: indeed, for $\mu := \sum_{i=1}^n P_i \in (\mathcal{L}')_+$ with $P_i \in \mathcal{P}$ for $1 \leq i \leq n$, choose $F_\mu \in L^\infty_{B*}(\mu, B)$ as in $(*)$, then pick versions $\psi_i \in X$ of the Radon/Nikodym densities $\frac{dP_i}{d\mu}$, and elements $\Psi_i \in F_{P_i}$, $1 \leq i \leq n$. Because of

$$\int \varphi \cdot \langle x, \Psi \rangle \, dP_i = \int \langle x, \varphi \cdot \Psi \rangle \, dP_i = \int \langle x, \varphi \cdot \Psi \rangle \cdot \psi_i \, d\mu$$
$$= \langle x, \int (\varphi \cdot \psi_i) \cdot \Psi \, d\mu \rangle = \langle x, ((\varphi \cdot \psi_i)^{\cdot} \cdot F)(\mu) \rangle$$
$$= \langle x, (\dot{\varphi} \cdot F)(\dot{\psi}_i \cdot \mu) \rangle = \langle x, (\dot{\varphi} \cdot F)(P_i) \rangle$$
$$= \langle x, \int \varphi \cdot \Psi_i \, dP_i \rangle = \int \varphi \cdot \langle x, \Psi_i \rangle \, dP_i \quad \text{for all} \ \varphi \in X$$

we then conclude

$$\langle x, \Psi_i \rangle = \langle x, \Psi \rangle \ P_i\text{-almost surely for all} \ i \in \{1, \ldots, n\}$$

for all $x \in B$ and, by consequence, infer

$$\Psi \in \bigcap_{i=1}^n F_{P_i},$$

showing even the relation $F_\mu \subseteq \bigcap_{i=1}^n F_{P_i}$ to be true. By construction of the maps F_P, $P \in \mathcal{P}$, in part (b), we have

$$\|\tilde{F}\|_{\mathcal{P}} \leq \|\|F\|\|.$$

104

(c) If $\tilde{F} \in {}^{B}\tilde{\mathcal{M}}$ is given and if $F \in {}^{B}\mathcal{M}$ is constructed as in part (a) of the proof, then for all $P \in \mathcal{P}$,

$$(\dot{\varphi} \cdot F)(P) = F(\mu) = \int \Psi \, d\mu = \int (\varphi \cdot \Psi) \, dP \quad \text{for all } \Psi \in F_P,$$

where $\mu := \dot{\varphi} \cdot P$ fulfills $\mu \in \mathcal{L}'$ because of $|\mu| \le \|\varphi\|_{\infty} P$. Hence condition $(*)$ in part (b) is satisfied. Conversely, assume that $F \in {}^{B}\mathcal{M}$ is given and that $\tilde{F} \in {}^{B}\tilde{\mathcal{M}}$ is constructed as in part (b). Let $\mu \in \mathcal{L}'$ fulfill $|\mu| \le Q := \sum_{i=1}^{n} \alpha_i P_i$, where $\alpha_i \ge 0$ and $P_i \in \mathcal{P}$ for $1 \le i \le n$. Choose $\Psi \in \bigcap_{i=1}^{n} F_{P_i}$, and a version $\varphi \in X$ of the *Radon/Nikodym* density $\frac{d\mu}{dQ}$. Then, according to condition $(*)$ in (b),

$$\int \Psi \, d\mu = \sum_{i=1}^{n} \alpha_i \int \varphi \cdot \Psi \, dP_i$$

$$= \sum_{i=1}^{n} \alpha_i (\dot{\varphi} \cdot F)(P_i) = F(\mu).$$

(d) If $B = \mathbb{R}$ holds, then the considerations in part (b) show with the help of Proposition 1.5(a) that for any $\varphi \in X_+$ and for any $f \in \mathcal{M}$ and $g \in \mathcal{M}$ satisfying $f \le g$

$$\int \psi_P \cdot \varphi \, dP = (\dot{\varphi} \cdot f)(P) \le (\dot{\varphi} \cdot g)(P)$$

$$= \int \psi'_P \cdot \varphi \, dP \quad \text{for all } P \in \mathcal{P}$$

holds whenever $\psi_P \in f_P$ and $\psi'_P \in g_P$ are arbitrary representatives of their classes. Hence $\tilde{f} \le \tilde{g}$ follows. Conversely, the relation $\tilde{f} \le \tilde{g}$ yields, by part (a),

$$f(\mu) = f'(\mu) = \int \psi \, d\mu = \sum_{i=1}^{n} \alpha_i \int \psi \cdot \varphi \, dP_i \le \sum_{i=1}^{n} \alpha_i \int \psi' \cdot \varphi \, dP_i$$

$$\le \sum_{i=1}^{n} \alpha_i \int \psi' \cdot \varphi \, dP_i = \int \psi' \, d\mu = g'(\mu) = g(\mu),$$

for all $\mu \in (\mathcal{L}')_+$ and all $\psi \in \bigcap_{i=1}^{n} f_{P_i}$, all $\psi' \in \bigcap_{i=1}^{n} g_{P_i}$, so that for continuity reasons we have $f \le g$. $\qquad \square$

Now we are able to show that the notion of coherence, as introduced in chapter 1, coincides with that defined in [*Hasegawa/Perlman 1974*] (cf. [*Siebert 1979*]):

Theorem A.2:

If $\mathcal{E} = (\Omega, \mathcal{F}, \mathcal{P})$ is an arbitrary statistical experiment then the following assertions are equivalent:

(a) $\mathcal{E}$ is coherent, which means $\dot{X} = \mathcal{M}$;

(b) for all $\tilde{f} = (f_P)_{P \in \mathcal{P}} \in \tilde{\mathcal{M}}$ there is a function $\psi \in X$ fulfilling $\psi \in \bigcap_{P \in \mathcal{P}} f_P$.

Proof: (b) $\Rightarrow$ (a): let $f \in \mathcal{M}$, and $\tilde{f} = (f_P)_{P \in \mathcal{P}} \in \tilde{\mathcal{M}}$ be as in part (b) of the proof of Theorem A.1. If $\psi \in f_P$ holds for all $P \in \mathcal{P}$, then part (c) of the proof of Theorem A.1 shows

$$f(\mu) = \int \psi \, d\mu = \dot{\varphi}(\mu) \quad \text{for all } \mu \in \mathcal{L}'.$$

A continuity argument (Proposition 1.1) yields the same equation to hold for all $\mu \in \mathcal{L}$. (a) $\Rightarrow$ (b): if $f = \dot{\psi}$ for some $\psi \in X$, then, according to $(*)$ in part (b) of the proof of Theorem A.1,

$$\int (\varphi \cdot \psi) \, dP = (\dot{\varphi} \cdot f)(P) = \int (\varphi \cdot \psi_P) \, dP \quad \text{for all } (\varphi, \psi_P) \in X \times f_P.$$

Hence $\psi = \psi_P$ holds P-almost surely and thus $\psi \in f_P$ for all $P \in \mathcal{P}$. $\qquad\square$

With the help of the preceding result we are now able to deal with coherence in the setting described in the remark before Theorem 5.14 in section 5.5.

Corollary A.3:

Assume that the experiment $\mathcal{E}$ satisfies the conditions specified in the remark preceding Theorem 5.14. Furthermore, assume that for all $\tilde{f} = (f_P)_{P \in \mathcal{P}} \in \tilde{\mathcal{M}}$ there is a selection $\varphi_{X(\omega)} \in f_{P_{X(\omega)}}$, $\omega \in \Omega$, such that the map

$$\varphi_f : \omega \mapsto \varphi_{X(\omega)}(\omega)$$

from Ω to $\mathbb{R}$ is $\mathcal{F}$-measurable. Then $f = \dot{\varphi}_f$ holds, and $\mathcal{E}$ is thus coherent.

Proof: Let $\mu \in \mathcal{L}'$. To be more specific, assume that $|\mu| \leq Q := \sum_{i=1}^{n} \alpha_i P_{X(\omega_i)}$ holds, where $\alpha_i \geq 0$ and $\omega_i \in \Omega$ for $1 \leq i \leq n$. Denote by $\psi \in X$ a version of the *Radon/Nikodym* density $\frac{d\mu}{dQ}$. By consequence (cf. the proof of Theorem A.1),

$$\dot{\varphi}_f(\mu) = \sum_{i=1}^{n} \alpha_i \int \varphi_f \cdot \psi \, dP_{\chi(\omega_i)}$$

$$= \sum_{i=1}^{n} \alpha_i \int_{\{\omega \in \Omega : \chi(\omega) = \chi(\omega_i)\}} \varphi_{\chi(\omega_i)} \cdot \psi \, dP_{\chi(\omega_i)}$$

$$= \sum_{i=1}^{n} \alpha_i f(\dot{\psi} \cdot P_{\chi(\omega_i)}) = f(\dot{\psi} \cdot Q) = f(\mu),$$

implying via Proposition 1.1 the claimed assertion. $\qquad\square$

Torgersen's [1980] approach has the advantage that unbounded random elements can be generalized in an obvious way: let $L_B^\infty(\mu)$ denote the set of $\overset{\mu}{\sim}$-equivalence classes F_μ of *Bochner* integrable random elements $\Phi : \Omega \to B$ fulfilling

$$|||\Phi|||_{\infty,\mu} := \inf\{\alpha \geq 0 : |||\Phi(\omega)||| \leq \alpha \ \mu\text{-almost surely }\} < +\infty,$$

and equip it with the norm given by

$$\|F_\mu\|_{\infty,\mu} := |||\Phi|||_{\infty,\mu},$$

where $\Phi \in F_\mu$ is an arbitrary representative of its class. Then $L_B^\infty(\mu)$ is a *Banach* space. Now for $1 \leq p \leq +\infty$, the set

$$\mathcal{L}_B^p := \{\tilde{F} = (F_P)_{P \in \mathcal{P}} \in \bigotimes_{P \in \mathcal{P}} L_B^p(P) : \bigcap_{P \in \mathcal{Q}} F_P \neq \emptyset \text{ for all } \mathcal{Q} \in \Xi(\mathcal{P})\}$$

generalizes the set of all random elements in B which are p–fold integrable with respect to all $P \in \mathcal{P}$, Evidently $\mathcal{L}_B^p$ is a locally convex topological vector space, the uniformity of which is given by the system of seminorms defined by $\|\tilde{F}\|_{p,P} := \|F_P\|_{p,P}$, $P \in \mathcal{P}$. In general $\mathcal{L}_B^p$ is, however, not metrizable. The set

$$\dot{X}^B \simeq \bigcap_{p \geq 1} \{\tilde{F} = (F_P)_{P \in \mathcal{P}} \in \mathcal{L}_B^p : \bigcap_{P \in \mathcal{P}} F_P \cap X^B \neq \emptyset\}$$

is a sub-space of $\mathcal{L}_B^p$ dense with respect to this uniformity: indeed, choose for $\mathcal{Q} \in \Xi(\mathcal{P})$ and arbitrary $\tilde{F} \in \mathcal{L}_B^p$ a function $\Psi \in \bigcap_{P \in \mathcal{Q}} F_P$. Then

$$\Psi_{n,\mathcal{Q}} := 1_{\{\omega \in \Omega : |||\Psi(\omega)||| \leq n\}} \cdot \Psi \in X^B, n \in \mathbb{N},$$

are bounded random elements which fulfill

$$\int |||\Psi_{n,\mathcal{Q}} - \Psi|||^p \, dP \to 0 \quad \text{as } n \to \infty \text{ for all } P \in \mathcal{Q}.$$

Now choose $\tilde{F}_{n,\mathcal{Q}} = (F_{n,\mathcal{Q},P})_{P \in \mathcal{P}}$ such that $\Psi_{n,\mathcal{Q}} \in F_{n,\mathcal{Q},P}$ for all $P \in \mathcal{P}$. Then the net

$$(\tilde{F}_{n,\mathcal{Q}})_{(n,\mathcal{Q}) \in \mathbb{N} \times \Xi(\mathcal{P})} \text{ in } \dot{X}^B$$

approximates $\tilde{F}$. Thus one may view – as is indicated by *Torgersen* [1980] for the scalar case – $\mathcal{L}_B^p$ as completion of $\dot{X}^B$ with respect to the system of seminorms $(\|.\|_{p,P})_{P \in \mathcal{P}}$.

For $1 \le p \le q \le +\infty$ we clearly have $\mathcal{L}_B^q \subseteq \mathcal{L}_B^p$ as well as $\mathcal{L}_B^p \subseteq \mathcal{L}_{B**}^p$. If, furthermore B is reflexive or separable, then $L_{B*}^\infty(P,B) = L_{B*}^\infty(P)$ holds [*Ionescu-Tulcea/Ionescu-Tulcea* 1969, pp.79,93], which implies

$$^B\tilde{\mathcal{M}} \subseteq \mathcal{L}_{B*}^\infty.$$

Remark: To establish the last inclusion for general *Banach* spaces B, one has to consider, instead of $\mathcal{L}_{B*}^p$, the spaces

$$\mathcal{L}_{B*}^p(B) := \{\tilde{F} = (F_P)_{P \in \mathcal{P}} \in \bigotimes_{P \in \mathcal{P}} L_{B*}^p(P,B) : \bigcap_{P \in \mathcal{Q}} F_P \ne \emptyset \text{ for all } \mathcal{Q} \in \Xi(\mathcal{P})\},$$

$1 \le p \le +\infty$. By analogy to the transition from $L_{B*}^\infty(P)$ to $L_{B*}^\infty(P,B)$, it is easy to conceive that $L_{B*}^p(P,B)$ consists of $\overset{P}{\sim}*$-equivalence classes of maps $\Psi : \Omega \to B^*$, such that

$$\langle x, \Psi \rangle \text{ is } \mathcal{F}\text{-measurable for all } x \in B \text{ and } \int^{\star} |||\Psi(\omega)|||^p \, P(d\omega) < +\infty,$$

$\int^{\star}$ denoting the upper integral with respect to P on $\mathbb{R}^\Omega$ in *Bourbaki*'s sense. The expression

$$\|F_P\|_{p,P} := \inf\{[\int^{\star} |||\Psi(\omega)|||^p \, P(d\omega)]^{1/p} : \Psi \in F_P\}$$

defines a norm on $L_{B*}^p(P,B)$, which spaces, again for reflexive or separable B, coincide with $L_{B*}^p(P)$ (see, e.g. [*Ionescu Tulcea/Ionescu-Tulcea* 1969, pp.17,80f,84,94]). $\triangle$

Some facts, for instance the relation of invariance or of $(\underline{S})$-optimality (concerning the scalar case $B = \mathbb{R}$, see again [*Torgersen* 1980]) to sufficiency, can be transferred easily from $\mathcal{M}^B$ to $\mathcal{L}_B^p$, thus translating some classical results treated, e.g. in [*Basu* 1970], [*Strasser*

1972], [*Schmetterer/Strasser* 1974], [*Schmetterer* 1977], or in [*Kozek* 1980], to the setting of generalized random elements. On the other hand, if we translate some other results like, for instance, the support criterion (Theorems 4.10 and 4.17) to $\mathcal{L}_B^p$, the outcome suffers from the essentially poorer topological-geometrical structure prevailing in non-metrizable topological vector spaces. This is – also in the classical context – the reason why the results specified in [*Bomze* 1986] are relevant primarily for dominated experiments.

A further, less favourable feature – for classical as well as for generalized random elements – is the "bifurcation" of optimality, ancillarity, and heredity notions in the case of unbounded random variables (cf. [*Torgersen* 1980] and, in the classical context, [*Bondesson* 1975] as well as [*Bahadur* 1979], where the phenomena described above explicitly are termed as "pathology of infinity"). This situation is also reflected in case of experiments consisting of boundedly complete, though incomplete, families of distributions as treated in [*Hoeffding* 1977a,b].

References

Bahadur R.R. [1957]: On unbiased estimates of uniformly minimal variance. Sankhyā **18**, 211–224.

Bahadur R.R. [1979]: A note on UMV estimates and ancillary statistics. In: *J.Jurecková* (ed.): Contributions to Statistics (*J.Hájek* memorial volume), 19–24. Reidel, Dordrecht.

Bahadur R.R./Lehmann E.L. [1955]: Two comments on "Sufficiency and statistical decision functions". Ann. Math. Statist. **26**, 139–142.

Basu D. [1970]: On sufficiency and invariance. In: *R.C.Bose* et al.(eds.): Essays in probability and statistics, 61–84. Univ. of North Carolina Press, Chapel Hill.

Basu D./Ghosh J.K. [1967]: Sufficient statistics in sampling from a finite universe. Proc. 36th Session Int. Statist. Inst., 850–859.

Bednarek-Kozek B./Kozek A. [1978]: Two examples of strictly convex non-universal loss functions. Preprint **133**, Institute of Mathematics, Polish Academy of Sciences, Warszawa.

Berger A. [1951]: Remarks on separable spaces of probability measures. Ann. Math. Statist. **22**, 119–120.

Blackwell D. [1947]: Conditional expectation and unbiased sequential estimation. Ann. Math. Statist. **18**,105–110.

Bomze I.M. [1984]: *Jensen*-Ungleichung in *Banach*-Räumen: ein kurzer Beweis. Anz. Österr. Akad. Wiss., math.-naturwiss. Klasse **121**, 67–68.

Bomze I.M. [1986]: Measurable supports, reducible spaces, and the structure of the optimal σ-field in unbiased estimation. Mh. Math. **101**, 27–38.

Bomze I.M. [1987]: Suffizienz, Invarianz und Optimalität in undominierten statistischen Experimenten I,II. Sitzungsber. Österr. Akad. Wiss., math.-naturwiss. Klasse **195**, 349–403 and 517–568.

Bondar I.V./Milnes P. [1981]: Amenability: a survey for statistical applications of the *Hunt/Stein* and related conditions on groups. Z. Wahrscheinlichkeitstheo. verw. Gebiete **57**, 103–128.

Bondesson L. [1975]: Uniformly minimum variance estimation in location parameter families. Ann. Statist. **3**, 637–660.

Burkholder D.L./Chow Y.S. [1961]: Iterates of the conditional expectation operators. Proc. Amer. Math. Soc. **12**, 490–495.

Diepenbrock F.R. [1971]: Charakterisierung einer allgemeineren Bedingung als Dominiertheit mit Hilfe von lokalisierbaren Maßen. Thesis, Univ. Münster.

Dixmier J. [1951]: Sur certains espaces considérés par *M.H.Stone*. Summa Brasil. Math. II, **11**, 151–181.

Dunford N./Schwartz J.T. [1964]: Linear operators. Part I. Wiley, New York.

Eberl W. jun.[1984]: On unbiased estimation with convex loss functions. In: *E.J.Dudewicz/D.Plachky/P.K.Sen*(eds.): Recent results in estimation theory and related topics, 177–192. Oldenbourg, München.

Fell J.M.G. [1956]: A note on abstract measure. Pacific J. Math. **6**, 43–45.

Fisher R.A. [1922]: On the mathematical foundations of theoretical statistics. Philos. Trans. Roy. Soc. London, Ser. A **222**, 309–368.

Fisher R.A. [1925]: Theory of statistical estimation. Proc. Cambridge Philos. Soc. **22**, 700–725.

Fremlin D.H. [1978]: Decomposable measure spaces. Z. Wahrscheinlichkeitstheorie verw. Geb. **45**, 159–167.

Halmos P.R./Savage L.J. [1949]: Application of the *Radon/Nikodym* theorem to the theory of sufficient statistics. Ann. Math. Statist. **20**, 225–241.

Hasegawa M./Perlman M.D. [1974]: On the existence of a minimal sufficient subfield. Ann. Statist. **2**, 1049–1055.

Hasegawa M./Perlman M.D. [1975]: Correction to "On the existence of a minimal sufficient subfield". Ann. Statist. **3**, 1371–1372.

Heyer H. [1982]: Theory of statistical experiments. Springer, Berlin.

Hirzebruch F./Scharlau W. [1971]: Einführung in die Funktionalanalysis. BI–Wissenschaftsverlag, Mannheim.

Hoeffding W. [1977a]: Some incomplete and boundedly complete families of distributions. Ann. Statist. **5**, 278–291.

Hoeffding W. [1977b]: More on incomplete and boundedly complete families of distributions. In: *S.S.Gupta/D.S.Moore*(eds.): Statistical decision theory and related topics II, 157–164. Academic Press, New York.

Ionescu-Tulcea A./Ionescu-Tulcea C. [1969]: Topics in the theory of lifting. Springer, Berlin.

Kakutani S. [1941a]: Concrete representation of abstract L-spaces and the mean ergodic theorem. Ann. Math. **42**, 523–537.

Kakutani S. [1941b]: Concrete representation of abstract M-spaces. Ann. Math. **42**, 994–1024.

Klebanov L.B. [1972]: "Universal" loss functions and unbiased estimation (Russian). Dokl. Akad. Nauk SSSR **203**, 1249–1251.

Kozek A. [1980]: On two necessary σ-fields and on universal loss functions. Prob. and Math. Statist. **1**, 29–47.

Kozek A./Suchanecki Z. [1980]: Conditional expectation of selectors and *Jensen's* inequality. In: *W.Klonecki/A.Kozek/J.Rosiński*(eds.): Mathematical statistics and probability theory. Springer, New York.

Krein M./Krein S. [1940]: On an inner characteristic of the set of all continuous functions defined on a bicompact *Hausdorff* space (Russian). Dokl. Akad. Nauk SSSR **27**, 427–430.

Laurent P.J. [1972]: Approximation et optimisation. Hermann, Paris.

LeCam L. [1964]: Sufficiency and approximate sufficieny. Ann. Math. Statist. **35**, 1419–1455.

LeCam L. [1986]: Asymptotic methods in statistical decision theory. Springer, New York.

Lehmann E.L. [1959]: Testing statistical hypotheses. Wiley, New York.

Lehmann E.L./Scheffé H. [1950]: Completeness, similar regions, and unbiased estimation. Part I. Sankhyā **10**, 305–340.

Luschgy H. [1988]: Pairwise sufficiency and invariance. Osaka J. Math. **25**, 785–794.

Luschgy H./Mussmann D. [1986]: Products of majorized statistical experiments. Statistics and Decision 4, 321–335.

Luschgy H./Mussmann D./Yamada S. [1988]: Minimal L-space and *Halmos/Savage* criterion for majorized experiments. Osaka J. Math. **25**, 795–803.

Mussmann D. [1972]: Vergleich von Experimenten im schwach dominierten Fall. Z. Wahrscheinlichkeitstheorie verw. Geb. **24**, 295-308.

Neveu J. [1969]: Mathematische Grundlagen der Wahrscheinlichkeitstheorie. Oldenbourg, München.

112

Neyman J./Pearson E.S. [1936]: Sufficient statistics and uniformly most powerful tests of statistical hypotheses. Statist. Research Memor. **1**, 113–137.

Padmanabhan A.R. [1970]: Some results on minimum variance unbiased estimation. Sankhyā A **32**, 107–114.

Phelps R.R. [1983]: Convexity in *Banach* spaces: some recent results. In: *P.M.Gruber/ J.M.Wills*(eds.): Convexity and its applications, 277–295. Birkhäuser, Basel.

Pitcher T.S. [1965]: A more general property than domination for sets of probability measures. Pacific J. Math. **15**, 597–611.

Ramamoorthy R.V./Yamada S. [1983]: *Neyman* factorization for experiments admitting densities. Sankhyā A **45**, 168–180.

Rao C.R. [1952]: Some theorems of minimum variance estimation. Sankhyā **12**, 27–42.

Rogge L. [1972]: The relations between minimal sufficient statistics and minimal sufficient σ-fields. Z. Wahrscheinlichkeitstheorie verw. Geb. **23**, 208–215.

Rüschendorf L. [1987]: Unbiased estimation in nonparametric classes of distributions. Statistics and Decisions **5**, 89–104.

Sapozhnikov P.N. [1978]: Minimax and uniformly best unbiased estimates based on finite statistical structure with quadratic loss. SIAM Theor. Prob. Appl. **23**, 805–811.

Scalora F.S. [1961]: Abstract martingale convergence theorems. Pacific J. Math. **11**, 347–374.

Schachermayer W. [1976]: Zylindrische Maße und die *Radon/Nikodym*-Eigenschaft von Banach-Räumen. Thesis, Univ. Vienna.

Schaefer H.H. [1974]: *Banach* lattices and positive operators. Springer, Berlin.

Schmetterer L. [1974]: Introduction to mathematical statistics. Springer, Berlin.

Schmetterer L. [1977]: Einige Resultate aus der Theorie der erwartungstreuen Schätzungen. In: Trans. 7th Prague conf. information theory, statist. decision functions, random processes Vol. A, 489–503. Academia, Praha.

Schmetterer L./Strasser H. [1974]: Zur Theorie der erwartungstreuen Schätzfunktionen. Anz. Österr. Akad. Wiss., math.-naturwiss. Klasse **6**, 59–66.

Semadeni Z. [1971]: *Banach* spaces of continuous functions. Volume I. PWN, Warszawa.

Siebert E. [1979]: Pairwise sufficiency. Z. Wahrscheinlichkeitstheo. verw. Gebiete **46**, 237–246.

Strasser H. [1972]: Sufficiency and unbiased estimation. Metrika **19**, 98–114.

Strasser H. [1985]: Mathematical theory of statistics. de Gruyter, Berlin.

Tempel'man A.A. [1972]: Ergodic theorems for general dynamical systems. Transl. Moscow Math. Soc. **26**, 94–132.

To Ting On/Yip Wing Kai [1975]: A generalized *Jensen*'s inequality. Pacific J. Math. **58**, 255–259.

Torgersen E.N. [1980]: On complete sufficient statistics and uniformly minimum variance unbiased estimators. In: Teoria statistica delle decisioni. Sympos. Math. **25**, 137–153.

Vulikh B.Z. [1967]: Introduction to the theory of partially ordered spaces. Wolters-Noordhoff, Groningen.

Wald A. [1950]: Statistical decision functions. Wiley, New York.

List of symbols

$$1_A(x) = \begin{cases} 1 & \text{if } x \in A, \\ 0 & \text{otherwise.} \end{cases}$$